THE EVENT UNIVERSE

Crosscurrents

Exploring the development of European thought through engagements with the arts, humanities, social sciences and sciences

Series Editor
Christopher Watkin, Monash University

Editorial Advisory Board

Andrew Benjamin
Martin Crowley
Simon Critchley
Frederiek Depoortere
Oliver Feltham
Patrick ffrench
Christopher Fynsk
Kevin Hart
Emma Wilson

Visit the Crosscurrents website at www.euppublishing.com/series/cross

THE EVENT UNIVERSE

The Revisionary Metaphysics of Alfred North Whitehead

Leemon B. McHenry

EDINBURGH
University Press

© Leemon B. McHenry, 2015, 2020

Edinburgh University Press Ltd
The Tun – Holyrood Road
12(2f) Jackson's Entry
Edinburgh EH8 8PJ
www.euppublishing.com

First published in hardback by Edinburgh University Press 2015

Typeset in 10.5/13 Sabon by
Servis Filmsetting Ltd, Stockport, Cheshire
and printed and bound in Great Britain by
CPI Group (UK) Ltd, Croydon CR0 4YY

A CIP record for this book is available from the British Library

ISBN 978 1 4744 0034 3 (hardback)
ISBN 978 1 4744 7457 3 (paperback)
ISBN 978 1 4744 0035 0 (webready PDF)
ISBN 978 1 4744 0478 5 (epub)

The right of Leemon B. McHenry
to be identified as the author of this work
has been asserted in accordance with
the Copyright, Designs and Patents Act 1988,
and the Copyright and Related Rights
Regulations 2003 (SI No. 2498).

Contents

Preface

To explain all nature is too difficult a task for any one man or even
for any one age.
Tis much better to do a little with certainty & leave the rest for
others that come after you.
(Isaac Newton)

This is my shot at our most general theory of the world, that is, metaphysics. Unlike Newton, I cannot claim to have discovered any certainty for the view I advance, but even the incomparable Newton overestimated the epistemic status of his own achievements. Rather, I follow George Santayana's more humble and pragmatic view when he said: 'Here is one more system of philosophy. If the reader is tempted to smile, I can assure him that I smile with him . . .' (1955: v). The immensity of the metaphysical project has often been condemned by many philosophers and scientists as beyond human intelligence, yet with full knowledge of the difficulty, the impulse toward theoretical understanding continues from one generation to the next. Santayana smiles, Hume laughs and still there is the hope that the collective contributions of human genius amount to real progress.

For most of the twentieth and the twenty-first centuries, the dominant tradition in Anglo-American philosophy has been conceptual analysis. This approach has resulted in a view of philosophy as an increasingly specialised and purely *a priori* activity that is to be contrasted with the empirical researches of science. Physicists, biologists and chemists take on problems about the real world. What remains for philosophers is a sort of self-contained, puzzle-solving activity that is immune from any scientific influence or refutation. This has left philosophy in an impoverished state. It is no wonder that scientists often complain about the irrelevance of philosophy to the progress of scientific knowledge (Weinberg 1994: 166, Hawking 1993: 41). There are, of course, exceptions to the rule. Whitehead, Russell and Quine produced systems of philosophy that addressed fundamental questions about the physical world. They were mathematicians who kept abreast of physics and viewed philosophy as integral to the overall task of understanding the nature of reality. The theory advanced in the present work follows their lead in exploring problems on the border between science and metaphysics. The primary goal is a metaphysical theory that accommodates modern physics. As evidenced by the subtitle of this book, the work takes its inspiration from Whitehead's event theory but it is by no means merely a work of exposition. The point rather is to incorporate the combined insights of Whitehead, Russell, Quine and the pioneering theoretical physicists in exploring the advances of physics and cosmology. Indeed, if following Whitehead, our ontology is determined once we have established the conceptual scheme that best accommodates physics, then we will need to take account of the state of physics in the twenty-first century, especially with regard to progress towards a complete unified theory or the Theory of Everything.

The position I defend is *revisionary* because it overthrows our ordinary common-sense modes of thought; *naturalistic* because it begins to construct metaphysical principles from the natural sciences – physics and cosmology in particular; and *realistic* because its naturalism demands the scientist's robust sense of a mind-independent reality as a foundation for enquiry into the nature of the physical world. In this regard, the influence of Whitehead, Russell and Quine is obvious. The influence of Whitehead is the greatest because he contributed the pioneering ideas in developing the event ontology. He also saw early in the twentieth century that unification is the name of the game in physics. As his thought developed from the philosophy of physics to metaphysics, his quest for unification broadened as he sought the ultimate generalities. But as with the position I espouse in this work, Whitehead made no pretence to finality. As science evolves so does metaphysics. This is why, according to Whitehead's naturalism, philosophers seek the essence of the universe but they 'can never hope finally to formulate these metaphysical principles' (1978: 4). The same holds for science. At best we might have testable results demonstrating that our hypotheses are false, but not that they are true.

For most of this book, the metaphysics defended herein is limited by the fact that it excludes consideration about the nature of mind. It might be the case that some other metaphysical position expresses the ultimate nature of reality and serves as a solution to the "hard problem" of consciousness, but for my project here, namely, the ontological foundation of physics, like Whitehead in *The Concept of Nature* (1920) and *The Principles of Natural Knowledge* (1919), I have limited my focus to the nature of physical reality. The only exceptions are Chapters 6 and 7 where I briefly consider the role of consciousness in orthodox quantum theory and the philosophical implications of the event ontology for the mind-body problem and personal identity. I apologise in advance to the mathematical reader for the paucity of symbols in this book. Since this work is primarily philosophical in its orientation, I made no attempt to include a technical, equation-based treatment that is necessary in physics. The fundamental concepts are explained in elementary terms and where Whitehead's neologisms are introduced, I explain them in comprehensible terminology. My discussion of physical theories is restricted to their relevance to the event thesis.

This book has been long in the making but repeatedly interrupted by the demands of other work in the so-called 'real world'. Moreover, a scientifically-informed metaphysics can always be delayed until the science is adequately understood, which of course is endless as the process of science is endless. Preliminary work on this project has been published as 'Quine's pragmatic ontology' (1995a), 'Descriptive and revisionary theories of events' (1996), 'Quine and Whitehead: ontology and methodology' (1997), 'Maxwell's Field and Whitehead's Events' (2007), 'Extension and the theory of the physical universe' (2008), and 'The multiverse conjecture: Whitehead's cosmic epochs and contemporary cosmology' (2011). Quine published a critical reply to the third paper in the same issue of *Process Studies*, now twice anthologised. I thank the editors, John Stuhr of *The Journal of Speculative Philosophy*, Daniel Dombrowski of *Process Studies* and Rafael Hüntelmann of Ontos Verlag (now De Gruyter), for permission to republish here parts of these articles and chapters. In an open letter of 1995, Quine said of my proposal for this work: 'The ambitious project which he now envisages is of precisely the sort that I like to picture as the next flowering of philosophy and science: a merging of rigorous, logically sophisticated methodology and ontology with the physicists' findings and quandaries in cosmology and quantum mechanics.' Time will tell whether this book merits Quine's praise or accomplishes its goal.

The image on the cover of this book, Bubble Nebula (NGC 7635), taken by the Hubble Space Telescope captures the awe-inspiring beauty of our eventful universe. The Bubble Nebula is an expanding shell of glowing gas ten light-years across that surrounds a hot, massive star in our Milky Way galaxy. I thank the Space Telescope Science Institute (STScI), Baltimore, Maryland, for the use of this image.

I am grateful to the University of Edinburgh for granting me a visiting professorship in 2009, and to the American Philosophical Association and the Institute for Advanced Studies in the Humanities (IASH) in Edinburgh for providing me with two fellowships, one in 2009 and another in 2013, the second of which rekindled my enthusiasm for this project. The centre of the Scottish Enlightenment has its secure place in the history of philosophy and science. From David Hume, Adam Smith, Charles Darwin and James Clerk Maxwell in the eighteenth and nineteenth centuries to Whitehead's delivery of *Process and Reality* as the Gifford Lectures in 1928 and beyond, Edinburgh continues to be a source of inspiration.

I wish to thank in particular a number of philosophers and physicists who, over the years, have contributed to my understanding of the issues developed herein, either by positive influence or critical evaluation. Some of these in the first wave, now long perished but not forgotten, include: Timothy Sprigge, Victor Lowe, Dorothy Emmet and W. V. Quine, the latter three students of Whitehead at Harvard University in the 1920s and 30s. In fact, this work was originally conceived in 1986 at Johns Hopkins University when I was engaged by Victor Lowe to write two chapters on the philosophy of physics for Whitehead's intellectual biography. In the second wave, still very much with us, I express my gratitude to: Henry Stapp, George Allan, John Lango, Nicholas Maxwell, John Heil, Pierfrancesco Basile, Paul Sharkey, Christine Holmgren, Timothy Eastman, John Llewelyn and Pauline Phemister. In particular I thank Pierfrancesco Basile, Christine Holmgren, Timothy Eastman and John Lango for providing helpful suggestions to earlier drafts of the chapters. For other types of support and encouragement during this project, I thank Michael Baum, Skip Murgatroyd, Ronald McIntyre, Frank McGuinness, Ruth Addinall, John Cant, Dory Scaltsas, Andy Clarke, Anthea Taylor, Donald Ferguson, Tom Pedersen and Tracy Browne. I also thank my editors, Carol MacDonald of Edinburgh University Press and Christopher Watkin of Monash University, for seeing a place for this work in the interdisciplinary Crosscurrents series and for their assistance throughout the publication process.

I owe a special debt of gratitude to my long-time friend and colleague, Nicholas Maxwell, for three decades of philosophical sparring and for his extraordinary patience in guiding me through some of the most difficult concepts of theoretical physics. I have profited greatly from his insightful criticisms of the present work, and for this reason, this book is dedicated to him.

Leemon McHenry
Los Angeles, California 2014

Dedicated to Nicholas Maxwell,
good friend and fierce critic

1. *Introduction*

Φύσις κρύπτεσθαι φιλεῖ – Heraclitus

The real constitution of things is accustomed to hide itself.
(trans. G. S. Kirk)

A WORLD OF EVENTS

The expansion of the universe is an event, but so is the hurricane off
the coast of California, the traffic accident outside my window, and the
dance of subatomic particles in my cup of tea. So in addition to galaxies, bodies of land and sea, automobiles and cups of tea, there appear
to be activities, happenings or episodes; but do these form a distinct
class of entity that we identify as events? And if so, what is the status of
events in our general theory of the world? These are questions of ontology, the most basic theoretical enquiry into the nature of existence, and
the subject of the present work.

Some philosophers argue that events must be recognised along
with material substances. The tragic explosion of the space shuttle
Challenger, for example, is as much an entity as the space shuttle
Challenger itself or the astronauts. Others contend that only material
substances should be admitted to our ontology, while events have a
sort of secondary or dependent status.[1] In this view, the assassination
could not occur without there being in the first instance an assassin, a
victim and a bullet. Still others claim that people, places and things are
events too, even though they do not seem to fit our ordinary conception
of events. According to this view, events are not one kind of individual
that exists in a world with other kinds of individuals; they are the *only*
true individuals in the world. Physical objects in this view turn out to
be qualifications of events. The traffic accident is an event but so are

I

the car, the driver and the fly on the dashboard. Common sense tells us that this is stretching things a bit too far, but as we shall see, modern physics suggests otherwise.

Early in the twentieth century, three Cambridge philosophers, Alfred North Whitehead, Bertrand Russell and C. D. Broad, became champions of event ontologies that were thought to be compatible with emerging relativity theory.[2] At the mid-century mark, W. V. Quine, at the Cambridge on the other side of the Atlantic, developed this view for the same reason. For this group of theorists, events replaced Aristotelian substances as the primary constituents of the universe. As such, events are conceived as the basic units of space-time spreading throughout and overlapping within the electromagnetic field. As modern physics advanced, event ontologies gained further support from the un-thing-like behaviour of elementary particles described in quantum mechanics. Energy becomes the fundamental idea and the dematerialisation of nature implied by quantum mechanics required a new conception of the ultimate entities. So, from the largest scale of cosmological theory to the smallest scale of subatomic physics, it appeared to these philosophers that we needed a new foundation in ontology that embraces the most important concepts in advancing physics. According to the Cambridge philosophers, this task is accomplished by systematically interpreting all physical phenomena in terms of events.

With the widespread influence of logical positivism in the 1930s, however, mainstream philosophy and physics eschewed the traditional project of fundamental ontology. While philosophers laboured to provide a secure logical foundation for the sciences and demonstrate the meaninglessness of all metaphysics, quantum physicists adhered to the instrumentalist interpretation of the mathematical formalism, or the Copenhagen interpretation. Cool analysis was in; grand speculation was out. But as the project of logical positivism eventually wore down, metaphysics slowly gained acceptance once again in philosophy and it was no longer seen as disreputable to engage in speculation about ontological foundations. Even in physics, the search for the underlying reality of the mathematical formalism has led to competing quantum ontologies, all of which have developed in opposition to the orthodox Copenhagen interpretation. One of these theories synthesises elements of Whitehead and Heisenberg's views to develop a realist position that interprets quantum phenomena in terms of events.[3]

An ontology of events is not particularly new on the philosophical scene. From the Ancient Greek, Heraclitus, to twentieth-century thinkers such as Whitehead, Russell, Broad and Quine, the view has had a considerable following. Continental thinkers such as Friedrich

Nietzsche, Henri Bergson, Martin Heidegger and Buddhist philosophers have also proposed event theories, but for reasons very different from the Anglo-American philosophers. The common denominator lies in opposition to a bias in Western thought that has privileged, in one way or another, a view of the world as made up primarily of enduring substances.

In what follows I explore this bias in the Western philosophical tradition and provide criticisms of those philosophers who defend the primacy of material substance. In opposition to this orthodox tradition, I will challenge the very idea that making sense commits us to a substance ontology and argue that modern physics supports an event ontology. While the theories of special and general relativity have ruled out substance as the supporting ontology and quantum mechanics appears to lead to the same conclusion, it is still very much an open question as to the shape of a final, unified theory. But since physics is the most basic and comprehensive of all the physical sciences, enquiry into the nature of reality should begin here and not with pure armchair speculation or linguistic analysis.

DESCRIPTIVE VERSUS REVISIONARY APPROACHES

In the wake of logical positivism, philosophers were primarily concerned with an analysis of concepts and the functioning of language.[4] Peter Strawson is a significant contributor to this tradition with the approach that he defended in his *Individuals: An Essay in Descriptive Metaphysics*. As he says, 'Descriptive metaphysics is content to describe the actual structure of our thought about the world . . .' (1959: 9). It is not an attempt to determine the nature of reality but rather seeks to describe the basic concepts by which we normally think about the world. In *Analysis and Metaphysics*, he identifies the work of the analytical philosopher as a labour 'to produce a systematic account of the general *conceptual structure* of which our daily practice shows us to have a tacit and unconscious mastery' (1992: 7). In this approach, an analysis of ordinary language takes priority in the attempt to construct a metaphysical theory. Strawson himself finds strong historical links with the logic of Aristotle and the epistemology of Kant, both of whom contribute to what he calls 'our conceptual scheme'. Revisionary metaphysics, he claims, is concerned to produce a better structure of our thought about the world. He names Descartes, Leibniz and Berkeley as key examples of thinkers who have produced permanently interesting structures of revisionary metaphysics in the history of philosophy. And from what follows in *Individuals* there is little doubt that he would classify the four-dimensional world-view of Einstein-Minkowski or the event ontologies of Whitehead, Russell, Broad and Quine as clear cases of exotic or extravagant metaphysics in the revisionary style.[5]

In *Individuals*, Strawson says that revisionary metaphysics is at the service of descriptive metaphysics suggesting a priority of the latter over the former (1959: 9). But in a later essay, he draws a contrast between Quine's scientific and revisionary programme and his own and says: '. . . I am not concerned to evaluate the relative merits of these two conceptions. Each has its own worth and its own appeal; and the choice between them is, ultimately, perhaps, a matter of individual temperament' (1990: 312). Strawson prefers the analysis of concepts that remain as the indispensable core of thought in least refined and most sophisticated human beings, but he does not discount the idea that revisionary metaphysics is often the instrument of fruitful paths of conceptual change.

While espousing a form of descriptive or analytic metaphysics in the tradition of Strawson, Donald Davidson is less sympathetic with the very idea of alternative conceptual schemes envisioned by the revisionary metaphysician. In his defence of the priority of descriptive metaphysics against the revisionary approach of Quine, he writes:

> Like Quine, I am interested in how English and languages like it (i.e., all languages) work, but, unlike Quine, I am not concerned to improve on it or change it . . . I see the language of science not as a substitute for our present language, but as a suburb of it. Science can add mightily to our linguistic and conceptual resources, but it can't subtract much. I don't believe in alternative conceptual schemes, and so I attach a good deal of importance to whatever we can learn about how we put the world together from how we talk about it. I think that by and large how we put the world together is how it is put together, there being no way, error aside, to distinguish between these constructions. (1985: 172)

For descriptive metaphysicians such as Strawson and Davidson, the main point of metaphysics is to talk sense about worldly things both specifically and individually. This involves the attempt to establish which class of entities is basic or fundamental and which of the other classes are derived from the basic class.

SCIENCE AND PHILOSOPHY

The proper relationship between science and philosophy, or, more specifically, between science and metaphysics, is central to the distinction between descriptive and revisionary metaphysicians; the former view philosophical enquiry as a sort of self-contained activity of conceptual analysis immune to revision by science while the latter view metaphysics as the general end of theory on a continuum with science.[6] Descriptive metaphysicians such as Strawson and Davidson will therefore defend substance as a basic class of entity because they argue it is an indispensable part of the conceptual scheme of common sense. When Whitehead repudiated substance philosophy and posited an ontology of events in its place, many such as Strawson would claim that he ceased to make sense (1959: 46–7, 59–86).[7] If Whitehead is right, however, Maxwell, Einstein and Heisenberg, not ordinary language philosophers, are at the forefront of making sense. Revisionary metaphysicians reserve the right to overthrow our common-sense categories in the effort to construct a comprehensive, unified scheme that is consistent with advancing science.

Whitehead clearly thought that cosmology and metaphysics sought the general principles at the end of a continuum that begins with observation. Indeed, unlike the logical positivists or Karl Popper, he made no attempt to distinguish between science and metaphysics. The true method in philosophy, he argued, is like the flight of the airplane. It begins at the ground of particular observation, proceeds to flight in the thin air of imaginative generalisation and then returns to the ground for renewed observation rendered acute by rational interpretation ([1929] 1978: 5). In this regard, his view of the correct procedure for the construction of metaphysical principles approximates the traditional method of scientific enquiry, that is, observation, hypothesis formulation, and drawing out test implications of the hypothesis. So, according to Whitehead, a system of metaphysics is meant to be judged in terms of whether the generalities pass the test of application, whether they illuminate our experience of the world, have broad explanatory power and provide unifying concepts for the sciences. In opposition to pure, *a priori* metaphysics, he argued that:

> Metaphysical categories are not dogmatic statements of the obvious; they are tentative formulations of the ultimate generalities.
>
> If we consider any scheme of philosophic categories as one complex assertion, and apply it to the logician's alternative, true or false, the answer must be that the scheme is false. The same answer must be given to a like question respecting the existing formulated principles of any science. (ibid.: 8)

Metaphysical megalomania in the likes of Descartes, Spinoza, Hegel and Bradley is thereby cured by a naturalised approach inspired by the American pragmatists, Pierce, James and Dewey. The quest for certainty is abandoned in both philosophy and science. In fact, Whitehead says 'even in mathematics the statement of the ultimate logical principles is beset with difficulties, as yet insuperable' (ibid.). The reference of this statement is his monumental work with Russell on the foundations of mathematics, *Principia Mathematica* (1910–1913), thought to be the last frontier in certainty but even this project was upset by Kurt Gödel's incompleteness theorem in 1931.

For Whitehead, the metaphysician attempts to establish a foundation for the unity of the special sciences by proposing a set of working hypotheses for their coordination. These function as ultimate generalities from which the metaphysician now seeks to discover whether they are applicable beyond their limited origin. Science gains from metaphysics the systematic overview of the fundamental concepts lying behind the specialised modes of enquiry, and metaphysics gains from the special sciences the empirical discovery of the special features of order in the universe. In this manner, the metaphysician's generalities are derived from empirical science, yet they push well beyond any particular science or sub-specialisation within science in virtue of the attempt to present a coherent set of categories for all the sciences. Whitehead's event theory and his later process metaphysics were advanced for this very reason.

When Russell took up the problem of our knowledge of the external world, he proposed a new method whereby philosophy might look to physics for achieving precise results. This is not to say that the results of physics give us certain truth, for if the revolution of modern physics has taught us anything, it is to be humble in our expectations. Nonetheless, physics still stands out as the discipline that philosophy might try to emulate. The so-called 'truth of physics' he accepts not as final, dogmatic statements of authority, because, of course, no one supposes the physics of one's day is infallible or incapable of improvement, but rather because a philosopher who wishes to know about the nature of reality would do well 'to assume whatever the consensus of physicists advises us to assume' (1946: 701). Those who have failed to heed this advice and set up a philosophy against physics have always ended in disaster. Even when considering those doctrines of physics most subject to doubt, we should remind ourselves that they are the best that the human intellect has achieved. Armed with the mathematical logic that he and Whitehead helped to create, Russell argued that philosophy should attempt to secure answers to smaller problems, and only on that basis proceed to larger generalities. But crucially it cannot ignore science in the manner in which he believes the idealists were guilty of doing. Similarly, with the rise of linguistic philosophy in the twentieth century, Russell with characteristic wit quipped that the later Wittgenstein 'seems to have grown tired of serious thinking and to have invented a doctrine which would make such an activity unnecessary' ([1959] 1995: 161). While Russell was firmly committed to understanding the world, Wittgenstein and his followers have told us 'it is not the world that we are trying to understand, but only sentences, and it is assumed that all sentences can count as true except those uttered by philosophers' (ibid.). Russell says this is perhaps an overstatement, but not by much. Linguistic philosophy, for him, was just the latest manifestation of a trend that made ignorance of science a virtue. Philosophy, under such restraints, was in danger of becoming idle amusement, like parlour games for the upper classes, and therefore irrelevant to fundamental problems.

Quine claims a direct influence from Russell in his construction of a scientific philosophy but his view of metaphysics as a generalisation of science is closer to Whitehead's view. He did, however, react to a different philosophical milieu of the mid-twentieth century when he resurrected this new form of metaphysics from the hands of the logical positivists.[8] As he came to recognise the failure of such projects as Rudolf Carnap's attempt to reduce all theoretical language of science to experimental terms, he became increasingly committed to a holism in which the theoretical and experiential terms of a scientific theory work together when confronting experience. The positivists who sought to separate these two had thrown the baby out with the bathwater because scientific theories far exceed strict observation and are therefore just as unverifiable as metaphysical theories. Even observation loses its special significance with the realisation that our perceptions of data are coloured by assumptions that are deeply embedded in the theoretical framework.

Metaphysics, for Quine, should not be conceived as a transcendent or purely speculative enterprise of the rationalists, but rather as a discipline that originates in the natural sciences and forms an essential part of our general theory of the world. It must secure its relevance within science; it is fallible, tentative and revisable like science and risks its dignity in the same manner by facing the tribunal of experience just like any scientific hypothesis (1981b: 72). In accordance with both Whitehead and Russell, he argues that metaphysics is naturalised when it is based on the best current account of the world that science has to offer.[9] The ontology posited by the philosopher and the ontology posited by the scientist both originate in experience. They differ only in their respective generality. Since physics is the science *par excellence* in virtue of its place in the hierarchy of the sciences, we begin to construct a generalised ontology from the posits of physics. As Quine puts it: 'our ontology is determined once we have fixed upon the over-all conceptual

scheme which is to accommodate science in the broadest sense ...'
(1953: 16–17). We cannot, he argues, rise above our scientific theories,
and, in particular, our theories of physics.[10]

For Quine, the descriptive metaphysician does not begin to do justice
to the role of science in shaping our world-view. In fact, it appears to
sidestep science in the attempt to discover another way to truth by
restricting philosophy to mere conceptual analysis. Robert Barrett and
Roger Gibson make this point:

> ... in the present day, when physics in particular stands as the monumentally
> impressive source of knowledge about the world that it does, philosophy can
> easily seem a laughable anachronism if its practitioners pretend to *another
> way* that dispenses with all that scientific stuff – experiment, measurement,
> controlled observation and theoretical teamwork – in favor of sheer reason-
> ing (or whatever else). (1990: xvii)

Instead, philosophy must work with cutting-edge discoveries of science
in the pursuit of truth. If physics posits an ontology that is contrary
to common sense and ordinary language, then so much the worse
for common sense. The revision of the categories is always a genuine
option for philosophy and science. In this way, the space-time regions
of the energy field or events might turn out to be ontologically basic
rather than the macro-bodies of sense perception.

For the revisionary, naturalised metaphysician, the relation between
the sciences and metaphysics is collaboration rather than contrast.
Conceived in this manner, metaphysics as 'first philosophy' must sur-
render its traditional claim to a truth beyond the empirical realm of
scientific investigation. A plea for open systems replaces the alleged
finality of absolute principles or the sacrosanct status of the common-
sense conceptual scheme.

THE QUEST FOR A UNIFIED THEORY

Metaphysics as conceived by Whitehead, Russell and Quine is closely
related to a fundamental project of theoretical physics, namely the
quest for unification of physical theory. Some of the greatest advances
in physics were achieved when previously believed separate phenomena
were discovered to be aspects of the same phenomenon. Celestial
and terrestrial motions were united in Newton's law of gravitation.
Electricity, magnetism and light were unified in the electromagnetic
theories of Faraday and Maxwell. Mass and energy were fused in
Einstein's special theory of relativity. Space and time were unified in
his special and general theories of relativity. More recently, Weinberg,
Glashow and Salam have extended Maxwell's theory by unifying the

electromagnetic and weak nuclear forces in what is now called 'the electroweak force'. Without a doubt, unified theories offer theoretical simplification achieved in the form of elegant equations and broad explanatory and predictive power. They are the central focus of theoretical physics and scientific progress is often measured in terms of success toward this goal.

Newtonian mechanics achieved one of the longest-lasting episodes of unification in physics. But beginning with Maxwell's electromagnetic theory, Einstein's theory of relativity and quantum mechanics, the seventeenth-century cosmology collapsed and the quest for unification in a new paradigm started all over again. The unification of general relativity and electromagnetism was, in fact, the focus of Einstein's later work, and while he did not achieve his goal of a unified field theory, his work nonetheless paved the way for the future direction of physics with a contemporary focus on grand unified theories and the Theory of Everything. To this day, however, the fundamental problem of unification of the electroweak force with the strong nuclear force in a grand unified theory – the electronuclear force – and the final link with general relativity remain unsolved. Simply put, we have a theory that describes how the planets orbit the sun and a theory that describes how electrons 'orbit' an atomic nucleus, but no theory that describes how these fundamental forces of nature are parts of one comprehensive theory. Assuming that the universe is comprehensible, it is a rational goal of physics to seek such a Theory of Everything. While Stephen Hawking was at one point optimistic about arriving at a final theory by the end of the twentieth century (1988: 156; 1993: 49), Michael Redhead claims that it is a receding horizon, compares such projects to the search for the Holy Grail and thanks the philosophers of science for keeping us down to earth with their more carefully-argued views (1995: 65, 86).

The unification of modern physics is an ontological problem because our theories of the large-scale structure of space-time and the small-scale, microphysical quantum phenomena are presently incompatible. The former expressed in general relativity is deterministic while the latter quantum theory (at least in standard, orthodox interpretations) is probabilistic. In fact, the indeterminism of quantum theory is exactly what prompted Einstein's famous remark to Niels Bohr about the dice-playing God. If we regard the standard model as the quantum theory of matter and its forces and general relativity as the theory of space-time, then the problem is to unify the theories of matter and space-time, perhaps in some form of a general field theory. One major source of the incompatibility concerns the very different roles that time plays in the two theories. Whereas general relativity is most naturally interpreted in terms of a four-dimensional view in which time is inseparable from space-time, orthodox quantum theory assumes the independence of time to the quantum system.[11] So the problem of how to conceptualise time in a consistent and comprehensive ontological theory is crucial. Physicists, however, with some notable exceptions, seldom think in the ontological categories of philosophers.[12] As we shall see in the following chapters, the positivistic resistance to ontology in our attempt to formulate a picture of physical reality, whether in relativity theory or quantum mechanics, has been a major source of failure to provide a comprehensive, unified theory.

Whitehead saw that the quest for a unified theory required rethinking the ontological foundations of physics and science more generally, but there are two distinct projects of unification that need to be distinguished: (1) the unification of physics, that is, the Big Problem of how to unify the known forces of nature – gravity, electromagneticism, weak and strong nuclear forces; and (2) the unification of the natural sciences in a general ontology. Whitehead's project was the second of these two; however, his unification project also applies to the Big Problem because a generalised ontology would address the problem of fragmentation of physical theory and provide a comprehensive, coherent view of reality. In other words, if we can make the metaphysical assumption that the natural universe itself is structurally unified, this unification will be mirrored first in an ontology of physics and then in the web-like interconnection in all the sciences.[13] This, of course, is no simple undertaking. The question of this book is whether Whitehead's proposal offers anything towards a solution to this problem, one that is technical and extremely difficult. Indeed, one might wonder how a theory produced early in the twentieth century can provide fruitful direction if the very best scientific minds working on the frontier of the unification project for well over seventy or eighty years have failed to solve the Big Problem. It might take eighty more years before we have our answer. I merely expect to explore the issues within the framework that Whitehead and others proposed. My aim in this book is therefore twofold: to defend an event ontology in the fashion of Whitehead, Russell and Quine against the traditional substance ontology, and to examine the plausibility of this theory within the contemporary quest for a unified theory of physics.

2. Descriptive Metaphysics

I assert that whenever a dispute has raged for any length of time, especially in philosophy, there was, at the bottom of it, never a problem about mere words, but always a genuine problem about things.

(Immanuel Kant)

As we have seen in Chapter 1, Strawson's famous distinction between descriptive and revisionary metaphysics describes two different methodological approaches to the question of fundamental ontology. In this chapter, I focus mainly on the origin of the descriptive enterprise in Aristotle's metaphysics and examine Whitehead and Quine's criticisms of this approach to the formulation of a viable conceptual scheme for modern physics. The project of descriptive metaphysics and the concept of substance are rejected on both scientific and philosophical grounds.

ARISTOTLE'S CONCEPTION OF SUBSTANCE

Aristotle spends considerable time in his *Categories* and *Metaphysics* working out the details of the nature of substance – a revival of the old Pre-Socratic question: what is ontologically basic?[1] He described his subject, metaphysics or first philosophy, as the study of 'being qua being', that is, the subject that concerns the issues that are most fundamental or at the highest level of generality. 'Being qua being' defines being simply as being, or what it is to be. The physicist studies one particular genus of being, but the philosopher's enquiry into the nature of being, claims Aristotle, is universal and deals with the nature of primary substance (1941: 736).

The basic categories of being are: substance, quality, quantity, time, place, position, state, relation, action and affection. All except the first have dependent being. For Aristotle, an analysis of ordinary language reveals these general categories and demonstrates how substance gains ontological priority over the other categories. The underlying supposition behind his metaphysics is: *grammar is the guide to ontology*. That is, he thinks that the way we speak about the world is no accident; the basic subject-predicate grammar faithfully corresponds to the way the world is really put together. This means that nouns and proper names identify substances, verbs and adverbs identify actions or events, and adjectives identify properties and class relations. Take, for example, the sentence: 'Space Shuttle *Challenger* exploded violently over the Atlantic Ocean off the coast of central Florida on 28 January 1986 at 11:38 EST.' Aristotle would say that 'Space Shuttle *Challenger*' identifies a particular concrete substance – that particular space shuttle, 'exploded violently' identifies an event and 'over the Atlantic Ocean off the coast of central Florida on 28 January 1986 at 11:38 EST' identifies a place and time. He contends that the introduction of nouns or proper names in the subjects of sentences typically identifies the primary reality of substance, which can be natural or artificial and complex entities.

So, Aristotle's procedure in the *Categories* is to deduce the categories from an analysis of linguistic forms. He proceeds by dividing language into 'things said without combination' and 'things said in combination', that is, simple terms (nouns and verbs) and prepositions and classifying the senses in which ordinary language predicates one term of another – hence the name 'categories' which means 'predications' – and thereby reaches a classification of being.

In the *Categories*, substance is characterised above all by its power to remain numerically one and the same while admitting contrary qualifications (ibid.: 13). So, most importantly:

1. Substances are self-same centres of change; that is, they endure throughout their changes as the same individual things. At one time, a ball may be hot; at another, cold. A woman may have blonde hair when she is younger and brown hair when she is older. These are contrary qualifications.

Aristotle further explains substance as follows:

2. Substance is not asserted of a subject, but rather is that of which everything else is asserted. That is, qualities are properly predicated of a subject – tabby cat, carnivore, mammal, animal; these are what are asserted or *said-of* a subject, not the substance, a primary individual thing. Substances are the subjects of which the other

categories do the job of predication. They are typically identified by nouns with indefinite articles – 'a man', 'a dog', 'a house' – or by proper names, 'Socrates', 'Fido', and so on. As the ultimate subjects of predication, which 'underlie everything else', singular individuals fulfil these conditions more adequately than species and genera. Aristotle specifically identifies this as the substratum. He notes further that substance is always subject and never predicate.

3. Substance is capable of independent existence. Qualities, on the other hand, are incapable of independent existence; that is, qualities exist only when attached to some particular individual thing. Similarly, words like 'to walk', 'to be healthy', 'to sit' do not imply self-subsistent things because the referents of these words are incapable of being separated from substance. That which walks or sits or is healthy is the existent thing. As he puts it in the *Metaphysics*:

> Now these are seen to be more real because there is something definite which underlies them (i.e., the substance or individual) which is implied in such a predicate; for we never use the word 'good' or 'sitting' without implying this. Clearly then it is in virtue of this category that each of the other also *is*. Therefore that which is primary, i.e., not in a qualified sense but without qualification, must be substance. (ibid.: 783)

4. Substance is a particular combination of form and matter. As opposed to Plato, Aristotle argued that forms do not exist independently of matter. And matter all by itself cannot give us a concrete, particular thing (ibid.: 785). Nothing exists without form. The matter and form give the substance a particular unity.

5. Substance has no contrary. There is no contrary of Socrates or a particular man. Qualities, on the other hand, have contraries – hot/cold, red/green, light/heavy, etc.

6. Substance does not admit of degree. Socrates might be more or less masculine but he cannot be more or less a concrete, individual man.

Although Aristotle does not explicitly take up the status of events, it is clear that he regards actions in the same manner as qualities; their existence is predicated on the basis of substances.

In addition to the analysis of substance in terms of form and matter, that is, *hylomorphism*, substances are also discussed in terms of their essential and accidental characteristics. Essential properties are those without which the substance could not continue to be that selfsame individual whereas the accidental properties can change without the destruction of the substance. As for essential properties, they are one and the same with the self-subsistent individual. All the categories have essences so that just as there is an essence to man or dog, there is an essence of white or musical; but the latter are secondary to the primary essences of substance, for as Aristotle puts it: 'primary things are those which do not imply the predication of one element in them of another element. Nothing, then, which is not a species of a genus will have an *essence* – only species will have it . . .' (ibid.: 787). Man or *Homo sapiens* then is the essence of an individual man, but pale man is not a species and is therefore an accidental quality.

In his *Physics*, Aristotle further explains the teleological nature of substance by way of his famous four causes – material, formal, efficient and final (ibid.: 240–1). His substances, and indeed the universe as a whole, are conceived in terms of purposeful activities. This is where the notion of organism comes into play in a particularly impressive way, as all substances are modelled on this notion. What a substance *is*, Aristotle called its actuality, and what it *can* become he called its potentiality. So, at one time a cat was actually a kitten and had the potentiality to become a fully-grown cat. Potentiality is the capacity of a substance to undergo a change of some kind, by either its own actions or those of other agents. To have a full explanation or account of a substance is to know its *telos* or final cause, which is the purpose of the substance. The final cause is the fulfilment of the formal cause, because once the substance has actualised its full potentiality it has developed fully its form or essence. The philosopher, for example, reaches perfection when he or she realises the form of rationality.

Substances can be regarded as existent beings apart from any changes that they undergo or as dynamic beings. The first is the logical approach whereby substances are classified by genus and species or investigated in terms of their substantial parts. The second is the teleological approach where substances are understood as dynamic processes of change. Either way, substances remain selfsame individuals.

The very idea of substance, of course, pre-dates Aristotle.[2] Pre-Socratic philosophers beginning with Thales sought to uncover the one stuff of the world. It was refined by the Pythagoreans in metaphysical conceptions of the soul as one that is simple and enduring throughout a person's life and surviving after the death of the body. Aristotle generalises the idea of substance as enduring for everything comprising natural processes.

STRAWSON AND DAVIDSON ON SUBSTANCE

Of the several characterisations of substance offered by Aristotle, two have exercised the most influence in descriptive approaches to metaphysics: (1) that which is capable of independent existence, and (2) that which is not asserted of a subject, but of which everything else is asserted. Strawson, for example, in his *Individuals* (1959), explicitly acknowledges the importance of these characterisations when he attempts to identify basic particulars, first in terms of ontological commitment via identifiability, and then in terms of his defence of two criteria for the subject-predicate distinction, the 'grammatical' criterion and the 'category' criterion. He says that he defends the 'traditional doctrine' that 'particulars can appear in discourse as subjects only, never predicates; whereas universals, or non-particulars generally, can appear either as subjects or as predicates' (ibid.: 137).

Following Aristotle, Strawson's investigation of 'our conceptual scheme' reveals that material bodies and persons are the basic particulars because they are paradigm logical subjects and only they satisfy the essential conditions of reference, namely, identification and re-identification. Identifiability establishes ontological commitment and identifiability-dependence establishes priority of ontological commitment. Hence, for Strawson, material bodies and persons are ontologically prior to other entities, for example, the private particulars of sense data or the unobservable entities of theoretical physics.

Given this scheme of thought, events turn out to be ontologically dependent since they are normally identified by the identification of material bodies. Births and deaths happen to particular creatures. Bangs, flashes, earthquakes, avalanches and battles are all products of material bodies in motion. The identification of events without reference to objects is problematic because, Strawson contends, events do not provide 'a *single*, comprehensive and continuously usable framework' of reference of the kind provided by physical objects (ibid.: 53). In other words, we would not be able to have ideas of births, deaths, battles and explosions without our ideas of persons and things interacting in various ways.

So, in accordance with the traditional doctrine of Aristotle and the medieval schoolmen, Strawson espouses a version of the old principle *operari sequitur esse*. Events and processes are subordinate to substantial things; they are all activities *of* bodies.[3] He also implies that a pure ontology of events would suffer from a serious practical problem of making sense of our talk about the world because we would not be able to make identifying references to anything. Re-identification becomes even more problematic in the case of events because we would not be able to re-identify such particulars as *the same again* without re-identification of material bodies.

Within the context of descriptive metaphysics, and indeed the mainstream of analytic philosophy, the question regarding the status of events focuses upon their possible admission to a substance-based ontology. Whether or not events gain the privileged status of 'particulars' along with primary substances depends essentially on an analysis of the logic of their grammar. Here we see clearly the Aristotelian doctrine – *grammar as a guide to ontology* – alive and well. As we have seen above, Strawson consistently argued against events as basic particulars. Davidson, on the other hand, has admitted events along with substances (or objects). Still others, 'anti-event metaphysicians', have driven the debate further by proposing to eliminate events completely.[4] The latter group is concerned mainly with ontological overpopulation; events for them add needless metaphysical baggage to our conceptual apparatus. They have argued that it is simply gratuitous to assume that events form an additional ontological category alongside of physical objects.

The question of the status of events entered the mainstream of contemporary philosophy in the 1980s as a result of Davidson's mind-body identity thesis and the production of actions formulated explicitly in terms of events. Much of this debate was focused on questions such as: 'Are there events?' and 'If so, then how are they identified and individuated?'[5] Advocates of the 'narrow view', such as Davidson, recognise events in addition to individual substances whereas advocates of the 'broad view', such as Whitehead, Russell and Quine, contend that events are the only concrete or basic particulars and substance is eliminated as an unnecessary substratum. The concern of the anti-event metaphysicians gets reversed; instead of eliminating events to keep our ontology tidy, we eliminate substances.

Events gain entry to Davidson's ontology because he believes that we cannot 'give a cogent account of action, of explanation, of causality, or of the relation between the mental and the physical, unless we accept events as individuals' (1980: 165). As mentioned above, the argument focuses on an analysis of ordinary language. If our ordinary ways of speaking call for a distinction between objects and events, then there is a need to recognise the metaphysics implicit in our language. Davidson writes:

> Our language encourages us in the thought that there are [events], by supplying not only appropriate singular terms, but the full apparatus of definite and indefinite articles, sortal predicates, counting, quantification, and identity-statements; all the machinery, it seems, of reference. If we take this grammar literally, if we accept these expressions and sentences as having the logical form they appear to have, then we are committed to an ontology of events as unrepeatable particulars . . . (ibid.: 181)

Davidson thus challenges Strawson's conception of ontological priority on the grounds that various grammatical structures and their logical forms designate event structures by identity and individuation. Events are named by gerunds – nouns generally ending in 'ing', for example 'the crashing of the jet at Los Angeles', or by verb-nominalisation – a noun or noun phrase formed from a verb, such as 'the crash of the jet at Los Angeles'. These phrases name particulars that have a verbal character; there is a verb alive and kicking, so to speak, in the description. They are often accompanied by adverbial clauses that modify the events that certain verbs introduce. In a slight modification of one of Davidson's famous examples, 'Sebastian walked slowly and aimlessly through the streets of Bologna at 2.00 a.m.', the reference clearly identifies an event – the slow and aimless walk of Sebastian (ibid.: 166–7). Davidson holds that the logical forms of sentences of this sort commit us to events. So, in accordance with Quine's criterion of ontological commitment encapsulated in his dictum, 'To be is to be the value of a bound variable', we render the walk of Sebastian in the language of first-order predicate logic as follows:

$\exists x$ (x is the walk and x is slow and x is aimless and x occurred in Bologna and x was at 2.00 a.m. and x was by Sebastian).

In this manner we quantify over events rather than objects.

Davidson therefore argues that there is no reason to assign second-rank status to events; while there is a conceptual dependence of the category of events on the category of objects, there is also a symmetrical dependence of the category of objects on the category of events (ibid.: 174). For example, we might identify an event by identifying an object involved in the event (the explosion is traced to a star that went supernova in 1987) and we might identity an object by identifying an event in which the object participates (the knife found in the airport dump is traced to the murder).

Strawson's famous distinction between descriptive and revisionary metaphysics provides a useful basis for the investigation of the different positions philosophers have held regarding the ontological status of events. The enormous amount of literature generated by these discussions has sharpened the focus and added clarity to event theory in the analytic genre, but it has completely overshadowed the reasoning behind the broader conception of events that has driven the revisionary endeavour.

WHITEHEAD AND QUINE: CRITIQUE OF ORDINARY LANGUAGE PHILOSOPHY

As opposed to the conceptual self-understanding of the descriptive metaphysician, the aim of the revisionary metaphysician is to understand the way the world is quite apart from the linguistic habits of common intercourse. Philosophers who pursue this general approach eschew the method of appealing to ordinary usage and refuse to restrict themselves to the categories already recognised in traditional metaphysics. This approach has gained strong support from modern physics since the revolutionary ideas have radically altered our ordinary, common-sense understanding of the world. In this regard, modern physics is a form of revisionary metaphysics that breaks out of the confinements of ordinary language. Given the theoretical latitude afforded by the revisionary approach to ontology, we are not bound by the straightjacket of one conceptual scheme, particularly the one that has become orthodox among ordinary-language philosophers. The Aristotelian or Kantian conceptual scheme is enshrined in our language, but not vindicated by it. It is indeed a peculiar feature of descriptive metaphysics that the universe is thought to be so amenable as to cooperate with the patterns of speech that developed in the Western world.[6] As Quine insists, however, we are not 'stuck with the conceptual scheme we grew up in' (1953: 78). And Broad says, with regard to the concept of the event introduced by relativity theory, that it 'is contrary to common usage, but common usage has nothing to recommend it in this matter' (1923: 54).

Our common-sense conceptual scheme has occupied a privileged place in Western philosophical thought. Whitehead, however, argued in *The Concept of Nature* that this resulted from the historical accident of the subject-predicate structure of Greek and then the dominance of Aristotelian logic which encouraged a conception of substance as the ultimate substratum not predicated of anything else (1920: 16–18).[7] Instead of consistently adhering to what is present in sense experience, the habit of postulating a substratum has reigned supreme in Western philosophy. Substance or matter, as the metaphysical substratum for the properties that are perceived, has become disconnected from the complex of immediate fact and survives only as an abstraction of thought. Later, in *Science and the Modern World*, Whitehead coins a new fallacy, 'misplaced concreteness' to characterise this mistake, namely, the error of mistaking the abstract for the concrete, or, more specifically, the error of attempting to build systems of thought based on abstractions rather than the truly real or concrete things (1925: 64). In particular, he believed that classical physics was guilty of misplaced concreteness since it had lost its empirical footing beginning with Galileo and Newton. While empirical in its methodology, classical physics was Aristotelian and Cartesian in its metaphysics for its adoption of non-empirical substances as the concrete real things at the base of nature and for its interpretation of these substances as enduring, inert matter subject to deterministic laws of mechanics.

Whitehead also argued against the associated epistemological doctrine, which he called 'the bifurcation of nature' (1920: 30). The doctrine holds that nature is partitioned into two systems of reality: a world of phenomenal appearances in the mind and a world of material objects that are the inferred causes of the appearances. According to this view, which was expounded in one form or another in the seventeenth century by Descartes and Locke and in the eighteenth century by Kant, the reality of the external world is never known but only conjectured, while the reality of the appearances is known but remains purely mental or dreamlike.[8] This view is also called 'the representational theory of perception'. Instead of perceiving nature as it really is, we 'suffer' poor representations of it. Whitehead argues that there is no way 'to establish any fundamental distinction between our ways of knowing about the two parts of nature as thus partitioned' (ibid.: 44). Science, however, is concerned with the coherence of the known, not with the cause of the known. So any view that bifurcates nature is a failure to make clear the relations between things perceptively known. The bifurcation of nature is the road to scepticism because once nature is partitioned into these two systems, it is difficult, if not impossible, to provide a coherent explanation for how our ideas correspond to an unknowable external reality. It was Hume who followed the logic of this argument to its ultimate end – empiricism devourers itself from inside ([1739] 1958).

Whitehead instead proposes to view nature as one system of relations. For him, there is no apparent nature, no 'psychic additions' as the causal result of an external reality. Nature is conceived as the field of experience rather than the cause of experience; what is known is immediate rather than mediated through the properties. In his own words, 'the red glow of the sunset should be as much part of nature as are the molecules and electric waves by which men of science would explain the phenomenon' (1920: 29). He further explains:

> The primary task of a philosophy of natural science is to elucidate the concept of nature, considered as one complex fact for knowledge, to exhibit the fundamental entities and the fundamental relations between entities in terms of which all laws of nature have to be stated, and to secure that the entities and the relations thus exhibited are adequate for the expression of all the relations between the entities which occur in nature. (ibid.: 46)

All knowledge of nature must come from within nature. Anything else is an artificial addition, for example, substance. The account of the coherence of things perceptively known is explained more fully in

Whitehead's metaphysics by the activities of present events that inherit the events of the immediate past (see Chapters 4 and 7). This theory replaces a line of thinking that began with Aristotle's hylomorphism and ended with the representational theory.[9]

The abstract concepts implicit in the structure of ordinary language have demonstrated their pragmatic value in enabling us to manage our common-sense world. In *Science and the Modern World*, for example, Whitehead says that the concepts of substance and quality that serve as the core of this common-sense view are 'the most natural ideas for the human mind. It is the way in which we think of things, and without these ways of thinking, we could not get our ideas straight for daily use' (1925: 66). And again in *Process and Reality* he writes:

> The simple notion of an enduring substance sustaining persistent quali-
> ties, either essentially or accidentally, expresses a useful abstract for many
> purposes of life. But whenever we try to use it as a fundamental statement
> of the nature of things, it proves itself mistaken ... For its employment in
> language and logic, there is – as stated above – a sound pragmatic defense.
> But in metaphysics the concept is sheer error. ([1929] 1978: 79)

Since Whitehead argued that our ordinary language fails to reflect the true nature of reality, he rewrote the metaphysical lexicon to capture the essence of process. The very concept of substance, deeply embedded as it is in our ordinary thought, hinders our ability to get at this dynamic nature of reality.

Ordinary language is designed to deal with our most practical needs, but not all sensed phenomena fall within its simplified classificatory scheme.[10] There is a plausible explanation for our survival as a species found in the fact that our senses are broadly adapted to tracking the world at the level of coarse macroscopic objects appropriate for everyday life, but this is not the whole story. Science, in fact, was considered an extension of our common-sense conceptual scheme until the ancient trinity of space, time and matter was replaced by the conception of the space-time energy field. The substance ontology only makes sense within a very narrow range of our current understanding of the electromagnetic spectrum – roughly within the range of visible light (400–700 nm or 3,800–7,200Å) – and even here there are the usual perplexing metaphysical questions about identity and individuation. That is, the traditional substance theory and Strawson's defence of a quasi-Aristotelian view privilege the sense receptors in human beings supplied by evolution, but in a way that presupposes that these receptors capture what there is in some more basic and universal sense. Our visual receptors are only sensitive to a narrow band of electromagnetic radiation within which entities sensed appear primarily as objects rather than events. Yet the field itself and the range of entities within it and perceived through it are predominantly of an event character rather than of an object character. Given the entire spectrum of known electromagnetic phenomena, from gamma rays to radio waves, the event theory is more plausible as a comprehensive ontology.

Present-day science and the foundational concepts of social intercourse thus remain unreconciled. On the one hand, we are aware that the advance of modern physics has replaced the seventeenth-century cosmology, yet, on the other hand, our common-sense notion of substance still reigns supreme in our ordinary thought. Our ordinary ways of sorting have been mistaken as a paradigm for determining what there is.

From Whitehead's point of view, another major problem with the descriptive account is that the kinds of events discussed by Strawson, Davidson and other descriptive metaphysicians are not the sort that would make an event ontology plausible. Strawson, for example, devoted a whole chapter of his book, *Individuals*, to sounds. His basic idea is that we could not make intelligible to ourselves a conceptual scheme in which material bodies were not basic, and in this hypothetical auditory world bodies are altogether absent (1959: 86). If this were meant to demonstrate the incoherence of an exclusive event ontology, it is misconceived as an argument against the likes of Whitehead, Russell and Quine. Moreover, the examples of battles, births, deaths, walks and airplane crashes are all every-day, garden variety events that necessarily involve references to objects. But physicists seem to have little trouble making sense of events and processes that are not tied down to physical particulars. For example, in the case of the cosmic microwave background, the microwave 'glow' that fills the sky is a remnant of the radiation that was present in the early universe that has now cooled to 2.7 K. Where is the physical object of which this radiation is a property? If we try to trace this event to an object, it appears that there is no presently existing object of which the radiation can be predicated. Moreover, this case is not analogous to that of the light from a star that ceased to exist a million years ago because there never was an object of which the event is a property. Instead, to the extent that the radiation can be traced to anything, it is to another event, namely the Big Bang, and this relationship is causal rather than one of substance to property. In short, because of the misguided notion that the evidence for events must be found in ordinary language, the event concept of descriptive metaphysics is too narrowly focused.

This brings us to Quine's criticism of descriptive metaphysics. While he was very much a central figure in the 'linguistic turn in philosophy' with his naturalistic-behaviouristic conception of language and key doctrines such as inscrutability of reference and indeterminacy of translation, Quine was also a systematic philosopher who attempted to answer the fundamental epistemological question: 'How do we acquire our theory of the world?'[11] That is, Quine constructs a system from the ontological ground-up beginning with the advances of science rather than viewing philosophy merely as a piecemeal, puzzle-solving conceptual activity. In this regard, he has more in common with other system-builders such as Whitehead and Russell than the likes of Ludwig Wittgenstein and J. L. Austin.

For Quine, ordinary language simply will not suffice when it comes to the role that science plays in acquiring our theory of the world. Descriptive metaphysicians therefore overdo the treatment of familiar material objects by regarding our ordinary speech as sacrosanct and by neglecting one of its most important characteristics, namely that it is always evolving. It is the job of philosophy and science to become clearer on things than ordinary speech typically permits and this is justification enough for scientific neologism (1960: 3–4). He therefore argued in accordance with Whitehead when he claimed that there is nothing special about the conceptual scheme of our ancestors, even as far back as *Homo javanensis,* but this does not mean any conceptual scheme is as good as any other (1953: 77). Alternative conceptual schemes or revisions to our conceptual scheme merit serious attention only if they provide fruitful results in science. This opens the door to an event ontology if indeed that theory is more suitable to modern physics. We always impose upon the world some conceptual scheme peculiar to our own language, but we can always improve upon our ordinary conceptual scheme as we explore possible advantages for scientific progress. Such revisions will occasion radical readjustments in the web of belief, but this is the price we pay for the pragmatic attitude towards experiment. When Quine argued in favour of conceptual scheme revision, he often referred to Otto Neurath's metaphor of the busy sailor who rebuilds his ship plank by plank while staying afloat on the open sea (1960: 3; 1981b: 72, 178). Our conceptual scheme, like the ship, must undergo constant revision. Quine writes:

> Our standard for appraising basic changes of conceptual scheme must be, not a realistic standard of correspondence to reality, but a pragmatic standard. Concepts are language, and the purpose of concepts and of language is efficacy in communication and in prediction. Such is the ultimate duty of language, science, and philosophy, and it is in relation to that duty that a conceptual scheme has finally to be appraised. (1953: 79)

While working within any one conceptual scheme, however, Quine treated any objects that are posited by theory as real. As he says: 'To call a posit a posit is not to patronize it . . . Everything to which we concede existence is a posit from the standpoint of a description of the theory-building process . . .' (1960: 22). Given that a theory has survived rigorous testing and has become part of an established theoretical framework, we adopt a moderate realism with respect to the objects posited. As Quine writes:

> For naturalistic philosophers such as I . . . physical objects are real, right down to the most hypothetical of particles, though this recognition of them is subject, like all science, to correction. I can hold this ontological line of naïve and unregenerate realism, and at the same time I can hail man as largely the author rather than discoverer of truth. I can hold both lines because the scientific truth about physical objects is still the *truth*, for all man's authorship. (1981a: 33–4)

Both pragmatism and realism are compatible in Quine's view because, as the authors of theories, we create truth, and, as believers of theories, we must view the theories as the truth about a reality external to us. Our choice of a conceptual scheme is pragmatic, but once we have made the choice, we treat the objects of the theoretical framework as ultimate – at least until evidence to the contrary requires radical adjustment.

Quine certainly shares this last point with C. I. Lewis who, in his pragmatic conception of the *a priori*, espoused the view that the *a priori* categories are necessary to any interpretation of experience, but they can always be challenged and changed subject to our attempts to improve our understanding of the world.[12] This pragmatic variation of Kant means that the categories are mutable by convention. Contrary to the descriptive metaphysician, the categories can be altered depending on how we choose to frame our concepts or define our words, but this does not mean that anything goes since the empirical facts remain unalterable to our will.[13] On this point, Quine remarks that the watchword is 'fallabilism' not 'relativism' (1981a: 34). If it is thought that there is any danger of relativism or an abandonment of truth that results from accepting the idea of an evolving conceptual scheme or alternatives to it, this charge does not apply to Quine. Some have understood him to be an ally of postmodern relativism, but this is a misrepresentation of his view. His fallabilism and naturalism commit him to the idea of an external reality, of which scientific testing determines whether the theories are worthy of our assent. Moreover, as a well-known proponent of holism and under-determination of theory by experiment, Quine contends that it is the whole theory, rather than separate statements, that face testing (1970: 5).

HAACK AND THE CONCEPTUAL
INVARIANCE THESIS

When Susan Haack took up the contrast between Whitehead and Strawson on descriptive and revisionary metaphysics, she claimed that she 'found it hard to resist reading Whitehead's work as a rather shrewd critique of Strawson's aims, methods and results' even though Whitehead's work appeared forty years before Strawson's (1979: 361). While her major objective was to clarify the main differences between Whitehead and Strawson, she ends up defending revisionary metaphysics and Whitehead's ontological priority based on observability of events over Strawson's criterion of identifiability of material bodies. Of particular interest is her discussion of Whitehead's rejection of what she calls 'the conceptual invariance thesis', namely the notion that there is a 'massive central core of concepts' that 'have one history' or, put another way, our conceptual scheme is common to people of different times and different languages. Strawson, for example, in his rather ambitious interpretation of his proposed view claims that there are no alternatives to our conceptual scheme. Whitehead's critique, however, implies that our ordinary conceptual scheme is neither privileged nor invariant; we are at liberty to substitute another more adequate for the purpose of advancing science, especially when we are seeking empirical foundations that are more consistent for the unification of the sciences.

Haack sees a deep-seated ambiguity in Strawson's views. His modest claim, which she calls 'the local connection assumption', asserts that different conceptual schemes are connected with different languages, and his ambitious claim, which she calls 'the global connection assumption', asserts that the same conceptual scheme is connected with all languages. The bridge from the modest claim to the ambitious claim results in the conceptual invariance thesis wherein Strawson claims 'there are categories and concepts, which in their most fundamental characters, change not at all' (1959: 10). If he holds to the local connection assumption, our conceptual scheme is merely one among others, but if he holds to the global connection assumption, there are no others.

Haack also points out that if Whitehead's thesis, that our conceptual scheme is a local, temporary accident inherited from the Greeks, is correct, then Strawson's global connection assumption must be false. She connects this thesis with linguist Benjamin Whorf's work on the American Indian languages, interestingly enough because one in particular, the Hopi language, has no subject-predicate distinction and carries a metaphysics in which events are basic (1970: 55–6). Similarly, Tsu-Lin Mei argues that Strawson is guilty of 'linguistic imperialism' because he takes English as the paradigm for all languages.[14] Since Chinese does not admit a distinction between subject and predicate, the so-called conceptual scheme of ordinary language describes at most 'a fact peculiar to Indo-European languages and has no further philosophical significance' (1961: 153). E. A. Burtt, who cites Whorf's work in his review of Strawson, comes to the same conclusion, namely, that our conceptual scheme might be much more changing, relative and culturally limited than Strawson assumes it to be (1953: 35). This is the principle of linguistic relativity in its strong form, sometimes called 'the Sapir-Whorf hypothesis', according to which the structure of a language affects the way its respective speakers conceptualise their world. Whorf's view is not that every language is associated with a different conceptual scheme, but rather that some languages are so radically different that they cannot be 'calibrated' and so they must be treated as being linked with quite different conceptual schemes (1970: 252). Whitehead, for example, would agree that Greek, Latin, French and English, all from the same linguistic family, share the same subject-predicate distinction and the same substance-property metaphysics and thus we should not be surprised that speakers of these languages conceptualise the world in the same way.

Haack thinks that Whitehead's view of the substance or the material-ist metaphysics relies on a weak, local connection assumption which asserts it is natural, though not compulsory, for speakers of a language to acknowledge a certain conceptual scheme (1979: 368). But now questions arise as to whether different languages can be translated into one another and whether there is a danger that any resulting transla-tion would always impose the conceptual scheme of the translator on the object language. The former would lend support to the global con-nection assumption and conceptual invariance, while the latter would result in a win for the local connection assumption and for conceptual scheme variation.

Haack provides a penetrating analysis of the issues and while she leans towards Whitehead's revisionary approach, she expresses reserva-tions about whether we can conclude that conceptual invariance is suf-ficiently doubtful because she makes no attempt to answer the question about translation (ibid.: 366).[15] But in my view this is not the central issue at stake here. It is not the question of whether one conceptual scheme can be translated into another thereby allegedly vindicating our ordinary concepts of everyday practice. The more crucial question is whether the world-view of modern physics actually provides support for the traditional ontology of substance. A similar point is made by Thomas Kuhn when he argues that the world-views postulated by scientists from one epoch to the next are incommensurable, that is, not inter-translatable (1962: 198). But whether paradigms are incommensurable or not has little bearing on the question of the appropriate underlying ontology. Scientists seem not to have any problem in shifting between the different world-views of classical and modern physics, all the while recognising that these are radically different pictures of reality.

HEIL ON LANGUAGE AND SUBSTANCE

While not a proponent of descriptive metaphysics, John Heil has argued for a substance-based ontology in his books, *From an Ontological Point of View* (2003) and *The Universe as We Find It* (2012). It is neither Aristotle nor Strawson's ontology of bodies that he advances but rather a view that is strongly influenced by Locke and Spinoza.

Honest philosophy, Heil argues, requires what the Australians call 'ontological seriousness'. So, with the same candour that he confronts the inescapability of the ontological project, he challenges contemporary philosophers to face up to the sterility of linguistic approaches and of the technically impressive, but ultimately unenlightening, contributions to philosophy. He therefore eschews the linguistic orientation that dominated descriptive or analytical metaphysics in the latter half of the twentieth century. What we need, he says, is something utterly different if we are to have any hope of advance and not simply 'more epicycles in going theories' (2003: 125). In arguing against the linguistic approach, he writes: 'If you start with language and try to work your way outwards, you will never get outside language. In that case, descriptions of the world, or "stories", go proxy for the world' (ibid.: viii). 'Genuine philosophical problems', he says, 'remain untouched by careful attention to ordinary language' (ibid.: 22). But what then is Heil's argument for substance if he rejects the linguistic methodology common to descriptive metaphysics?

Heil steers a course between reductionist doctrines that attempt to reduce human knowledge to an all-encompassing super-physics and a hierarchical conception of reality. He regards reductionism as a failure because it deprives us of crucial explanatory power and therefore turns out to be impractical. As he writes: 'Important higher-level patterns and relations are invisible to physics. The result is self-defeating and dehumanizing; we deprive ourselves of perspectives essential to an understanding of our place in the world' (ibid.: 31). But Heil also maintains that anti-reductionism does not imply a commitment to levels of reality. He rejects the idea that there are levels of reality (as opposed to levels of organisation or complexity) as 'a philosophical artifact spawned by a commitment to the Picture Theory' (ibid.: 73). This idea, he contends, is one that requires us to 'find an object, kind, or property correspond- ing to every significant predicate' (ibid.: 58), or, more simply put, it is the idea that language mirrors ontology. Heil finds problems with the Picture Theory insurmountable because it cannot account for causal relevance nor inter-level relations and as a consequence the hierarchical conception of reality fails as well.

Heil's positive thesis is a version of realism that recognises substances and properties, with an emphasis on the idea that properties are both qualitative and dispositional. This view is traced to Plato's Eleatic Stranger in *Sophist* who conceives of real things as nothing but powers. Locke immediately comes to mind as a modern expression of this notion. This is what Heil calls the 'identity theory' – that a property of a concrete object is simultaneously dispositional and qualitative. He writes:

> The identity theory is to be distinguished from theories according to which the dispositional and the qualitative are 'aspects', or 'sides', or higher-order properties of properties. A property's dispositionality and its qualitativity are, as Locke might have put it, the selfsame property differently considered. (ibid.: 112)

For example, the quality associated with an electron's mass is strictly identical with the power associated with the electron's mass, or an object's colour is strictly identical with its disposition to reflect light in a certain way.

When Heil turns his attention to objects or substances, once again we find that his analysis echoes Locke. Properties are found in objects; they are particularised ways that objects are. But what are objects exactly? Are they merely substrata for properties or thin particulars? Like Locke, Heil leaves it open as to whether or not what we ordinarily call 'objects' are objects in a more restricted sense. They might turn out to be thick- enings in the fabric of space-time or disturbances in energy fields. This raises the question of whether Heil's proposed ontology is a pluralism of objects and properties or monism with objects as modes. Heil leans toward the former but does not discount the possibility of the latter (ibid.: 41–2, 190). This important theme reappears in his next book.

In *The Universe as We Find It*, Heil claims that fundamental physics is in the business of telling us how the universe *is* whereas fundamental ontology tells us what the universe must be like if *any* theory is true (2012: 2). But if this is the case, it is unclear why substance, in his view, is the best ontological foundation for physics. In one crucial passage where Heil takes up 'the Big Picture', he rightly says:

Metaphysics developed alongside science. But science has come a long way since Aristotle. The science of Locke and Newton is not our science. Why should anyone imagine that categories that worked well enough for the Milesians, the medievals, and the philosophers of the Enlightenment would work in an era encompassing relativity and quantum mechanics? (ibid.: 11)

In what follows, however, there is really no discussion of why substance is required of relativity theory, quantum mechanics, or, more broadly, the standard model of particle physics. When Heil does take up quantum mechanics, the weirdness of quantum phenomena is attributed to bad metaphysics rather than the possibility that another ontology might be more suitable for the non-classical behaviour of subatomic particles (ibid.: 44–9). Likewise, when he discusses temporal parts of objects as space-time worms, he does not consider the possibility that objects in the four-dimensional manifold could be understood as event-sequences – as is standard in physics (ibid.: 35–6). Keith Campbell, for example, says in reference to this issue that once we realise that 'concrete particulars have temporal parts which are complete particulars in their own right, we soon travel out of the territory of concrete particulars altogether. A cosmology based on *things* gives way to one based on *events*' (1976: 33).

Heil says that the ontological picture he defends is neutral since it is up to physics to tell us specifically what there is, but he nevertheless insists that whatever it is, it will be simple, propertied substances. Science fills in the blanks left by the philosopher's general framework (2012: 41). He says his ontological picture is of a piece with conceptions implicit in Locke and Spinoza (2012: 11). This suggests that it is not ordinary bodies of perception such as a tomato which he uses repeatedly as his example of substance, but either Locke's corpuscles or Spinoza's One that satisfies the criterion that substances must be simple. This is required for them to be: (1) property bearers and (2) non-dependent. A substance is simple if it has no parts that are themselves substances. Since ordinary objects like the tomato do have complex substantial parts, they turn out to be Spinoza's modes.

Now when it comes to events, Heil argues that they fail the ontological test because they seem not to be property bearers, yet, like Strawson, the manner in which Heil conceives of events makes it difficult to understand how they could be property bearers. He says he is calling attention to 'a difficulty for the idea that an *event*, regarded as a substance's possessing a property at a time . . . could possess a property', but the example appears to assume that events are always dependent on substances. '*Thus characterized*', he says, events 'appear not to be apt property bearers' (2012: 22, my emphasis). This leaves it open as to whether there are other more coherent characterisations of how events might be property bearers, and perhaps Heil might be amenable to this idea. In the event ontology I have in mind, the event-object relation is reversed. Events are energetic activities interrelated in such a way as to form a network or field of relations; objects such as electrons, atoms, molecules and the like are disturbances or vibrations in the field. They are all patterns discernible in event-sequences. Thus conceived, as basic particulars, their role as property bearers is no more mysterious than the manner in which Aristotle and his followers understood substances to be property bearers. Moreover, events are also simple in the sense that the atomic, epochal occasions at the base of nature cannot be resolved into further events. What then is the main point of contention between substances and events? As we have seen above, Whitehead argued that substance, understood as a substratum for properties, does not serve science as an empirical foundation whereas events, as the ontological framework for objects, do provide this foundation.

If following Heil's own suggestion that the ontologist provides the broad conceptual framework within which the physicist, by detailed empirical investigation, fills in the details, then his simple substances in his bottom-up, scientifically-informed ontology must be either quantised or continuous, they must be either the corpuscles of particle physics or the entire space-time electromagnetic field (ibid.: 11, 278). What is unclear in his view is to what extent, if any, physics has the final authority about our current ontology. He says: 'The deep story about the universe's ontology is to be found in fundamental physics, but this does not augur the replacement of ontology by physics' (ibid.: 279). The ontology, according to Heil, must be scientifically informed but it cannot be reductionist. But if the 'deep story' is the business of fundamental physics and the physics of our time points to an event ontology, what objection has the philosopher from top-down?

There is much to be admired in Heil's focus on the importance of fundamental ontology and his analytically-rigorous approach to the subject, but in his defence of substance he ends up with results closer to descriptive metaphysicians irrespective of his rejection of their linguistic methodology. In light of my critical analysis of the weaknesses in the substance ontology, advocates of descriptive metaphysics may, nonetheless, still claim that an event ontology does not make sense. So, my task in what follows is to demonstrate not only that it does make sense but that modern physics requires it. Heil's view that ontology provides the broad conceptual framework for which physics fills in the details appears to be close to the view that I will defend in my later chapters, but with a major difference: the ontology that I will be defending is one that *arises out of physics* without imposing a substance framework where it has long outlived its purpose. The findings of Maxwell's electromagnetic theory, quantum mechanics and Einstein's theory of relativity imply an ontology of events rather than a substance ontology. That leaves us with the large-scale cosmological speculations about the universe, its origin, death and the possibility of its being merely one in a multiverse. While this part of my project is highly speculative, I show how the generally-accepted oscillation model generating *our* universe and others is perfectly consistent with a process, event ontology.

Heil remarks that it is unfortunate that we all labour under a multitude of assumptions, ignore apparent threats, talk past one another, operate in distinct worlds and only sometimes engage in meaningful discourse and debate (ibid.: 288). When we engage, as I hope to do, we might stubbornly hold on to our entrenched views, but at the very least there is clarification about the assumptions and whether one is reasonable in holding them.

3. The Influence of Modern Physics

> *What we observe is not nature itself, but nature exposed to our method of questioning.*
> (Werner Heisenberg)

Having exposed the weaknesses in the descriptive approach to ontology, I now take up the argument for a revisionary approach based on modern physics. In this chapter, I therefore investigate the influence of electromagneticism, relativity theory and the early quantum theory on the development of the event ontology in the 1920s with particular focus on Whitehead's view. These were the three key ideas that led to a transformation of our view of reality. The two main themes of this chapter include: (1) physical evidence in support of an ontology of events, and (2) the increasing unification of physical theory until we arrive at the current state of two highly successful, unified theories that are presently disunited within the search for a comprehensive, unified theory.

HISTORICAL PERSPECTIVE

The essence of Whitehead's theory of the physical universe is perhaps best grasped in contrast to the views expressed by the founders of the seventeenth-century cosmology, Galileo, Descartes and Newton, who advanced the doctrine of mechanistic materialism, the clock-like view of the universe as matter in motion. Newton provided the grand unification of terrestrial and celestial motions with his law of universal gravitation and his laws of motion and, in doing so, achieved a synthesis of everything known about the motions on earth and in the heavens. The centrepiece of this view is Newton's idea of the independence of space and time as defined in the Scholium of his *Philosophiae*

Naturalis Principia Mathematica ([1687] 1995). Objects in motion are understood within a three-dimensional space existing at different times. As Newton defined space in his *Principia*, he writes: 'Absolute space, in its own nature, without regard to anything external, remains always similar and immovable.' 'Absolute, true, mathematical time', he writes, 'of itself, and from its own nature flows equally without regard to anything external . . .' (ibid.: 13). As time advances from one instant to another, the whole universe is conceived to exist at once, complete and determinate. So, at a given moment of a person on earth, there is a simultaneous moment on Jupiter, Pluto, Alpha-Centauri and anywhere else in the cosmos, no matter how vast the distance. This is the concept of the cosmic now or absolute present that is central to Newton's notion of absolute simultaneity.

The concepts of absolute time and space imply that the temporal and spatial relations among objects are external. This means that the spatio-temporal relations have no effect on the essential nature of the things in space and time. On this score, Newton's view is sometimes called the 'billiard-ball model' of the universe. It is Aristotle's substance ontology reformulated for classical mechanics but with some important qualifications. In Aristotle's view, for instance, time is a measure of change. The very idea of time passing without changing objects would be incoherent for him whereas, for Newton, time exists 'without regard to anything external'. Time is completely independent and would carry on even if there were no events or objects. The same holds for space. Absolute time is required for precise mathematical measurements as a corrective to what he calls 'common time' as measured in hours, days, months and years. Absolute, 'immovable' space is required to give meaning to the absolute motion of a body independent of anything else in that space.

Both Galileo and Newton held variations of the old doctrine of atoms and the void, revived from the ancient Greek philosophers, Democritus and Epicurus. Space is a receptacle for matter in motion, they believed, since bodies can move only when there is a void that allows their passage by giving way and by offering no resistance. Newton makes his view clear in his '*De gravitatio et aequipondio fluidorum*' where he defines the terms 'place', 'body', 'rest', and 'motion' in opposition to Descartes' view, such that: (1) space is construed as being distinct from body, and (2) motion is determined with respect to the parts of space, instead of positions of contiguous bodies (1962: 121–56). An atom occupies a point of space, and in accordance with Newton's law of gravitation, all interaction is action at a distance.

Extension, for Newton, is understood to be neither substance nor accident; nor is it 'simply nothing' but 'it has a certain mode of existence proper to itself which suits neither substances nor accidents' (ibid.: 99). By contrast, Descartes had argued against the void by conceiving of the extended universe as one physical substance having the properties of motion, divisibility, mutability and flexibility. As Descartes makes the point in his *Principles of Philosophy*, he writes:

> As regards a vacuum in the philosophical sense of the word, i.e. a place in which there is no substance, it is evident that such cannot exist, because the extension of space or internal place, is not different from that of body. For, from the mere fact that a body is extended in length, breadth, or depth, we have reason to conclude that it is a substance, because it is absolutely inconceivable that nothing should possess extension, we ought to conclude also that the same is true of the space which is supposed to be void, i.e. that since there is in it extension, there is necessarily also substance. ([1644] 1911: 262)

In this way, space and bodies in space are thought of as varying solely in the degree to which they are thin or thick regions of extension rather than differing in kind. According to Descartes, space or extension is a plenum.

The mechanistic theory, together with advances in mathematics, seemed to be securely on the path to unlocking the secrets of nature. At almost every step of the way, all parts of the physical puzzle seemed to the scientists of succeeding generations to confirm Newton's grand cosmological scheme. With the revolution in physics that occurred with Einstein's special and general theories of relativity, however, it was clear that the whole foundation of physics needed rethinking. In the preface to *The Principle of Relativity*, wherein Whitehead presented his alternative rendering of the theory of relativity, he says that this theory 'takes its rise from that "awakening from dogmatic slumber" – to use Kant's phrase – which we owe to Einstein and Minkowski' (1922: v). The second scientific revolution in physics, he says in *Adventures of Ideas*, began with the wave theory of light and ended with the wave theory of matter (1933: 200).

Whitehead opposed the doctrine of mechanistic materialism on every point. First, as opposed to the mechanical conception of the universe, Whitehead's view of reality is more thoroughly organic (1925: 93). Second, in contrast to the absolute theory of space and time, he advanced a relative theory following Leibniz and Einstein. Third, in place of the primacy of instants of time and points in space, Whitehead argues that all perceptions of nature occur within durations; instants and points are derivative abstractions from perceptible durations. Fourth, Whitehead replaced the theory of substance or the materialist doctrine underlying the seventeenth-century cosmology with an ontology of events. Fifth, as Whitehead's theory evolves from his philosophy of physics to metaphysics, his understanding of events is refined such that the basic entities are conceived as atoms of experience. This, he claimed, is a repudiation of the doctrine of 'vacuous actuality', that is, mere inert matter ([1929] 1978: xiii).

MAXWELL'S ELECTROMAGNETIC FIELD

Whitehead considered James Clerk Maxwell's electromagnetic field to be the key idea for the development of the event ontology.[1] It was, in fact, the focus of Whitehead's earliest interests in mathematical physics beginning with his work conducted under the supervision of W. D. Niven at Cambridge in the 1880s.[2]

Einstein, who viewed Faraday and Maxwell's theory as the greatest change in the axiomatic basis of physics and in our conception of the structure of reality, identifies the course of the revolutionary idea as follows:

> ... under the pressure of observational facts the undulatory theory of light asserted itself. Light in empty space was conceived as a vibration of the ether, and it seemed idle to conceive of this in turn as a conglomeration of material points. Here for the first time partial differential equations appeared as the natural expression of the primary realities of physics. In a particular area of theoretical physics the continuous field appeared side by side with the material point as the representative of physical reality. This dualism has not to this day disappeared, disturbing as it must be to any systematic mind. (1982: 30)

If indeed Einstein is right about this dualism, it remains so despite the efforts of Whitehead. Although Whitehead postulated a dualism of events and properties (or 'objects' as he calls them), his conception of the uniform field of space-time eliminates the material objects or points. He says: 'The physical things which we term stars, planets, lumps of matter, molecules, electrons, protons, quanta of energy, are each to be conceived as modifications of conditions within space-time, extending throughout its whole range' (1933: 201–202). This approximates Descartes' concept of the plenum. The so-called empty space is, according to Whitehead's proposed interpretation, a uniform electromagnetic field which includes the vast regions of space between planets and galaxies, and the vast regions between subatomic particles in the atom.

In his *A Dynamical Theory of The Electromagnetic Field*, James Clerk Maxwell says: 'The theory which I propose may therefore be called a theory of the *Electromagnetic Field*, because it has to do with space in the neighbourhood of the electric and magnetic bodies, and it may be called a *Dynamical* Theory, because it assumes that in that space there is matter in motion, by which the observed electromagnetic phenomena are produced' ([1865] 1982: 34).

The major points of Maxwell's unification can be summarised as:

1. The development of a radically new entity expressed as the energy field.
2. Electricity and magnetism previously believed to be separate and distinct are unified into one entity, the electromagnetic field.
3. Light is unified under the more general theory of electromagnetism. Light is a relatively small band of electromagnetic waves within a vast spectrum of phenomena from radio waves to gamma radiation.
4. Electromagnetism and optics are thereby unified as electrodynamics. Light and electromagnetic waves have the same velocity and are manifestations of the same process.

The physico-mathematical concept of a field initially emerged from the work on fluid dynamics by Euler and others. It is represented as a continuum characterised by physical quantities that vary smoothly from place to place and moment to moment. In the theory of Faraday and Maxwell, the field is understood in terms of curved lines of force that are the sites of natural powers ready to act on material entities placed on them. Maxwell treated Faraday's electric and magnetic lines of force as a simple geometrical consequence of this disposition, present at any point in the surroundings of an electrified conductor or a magnetic body to point in a definite direction. This gives rise to the idea that the electric or magnetic force exerted on an object in the field is to be thought of as a vector at that point. The lines of force are the paths in space to which such vectors are tangent.

Faraday's pioneering experimental research resulted in the idea that the electric field that acts on electrically charged particles and the magnetic field that acts on magnets are actually one and the same field. In his attempt to explain the electromagnetic field, Maxwell produced elaborate equations in quaternion form that were later transformed into the elegant simplicity and symmetry of the vector calculus by Oliver Heaviside and William Gibbs. This theory is one of the great leaps of simplification in the history of physics. In Maxwell's system of partial differential equations, electric and magnetic fields appear as dependent variables. The electromagnetic field is always represented as a field of these two interrelated forces. When the equations are formulated for a vacuum having no charge and current, the interdependence becomes clear. A magnetic field changing in time creates a circulating electric field and an electric field changing in time creates a circulating magnetic field. In this way, the fields are no longer conceived to be distinct; a change in one creates a change in the other.

Once it was clear that electricity and magnetism are aspects of the same phenomenon further advances in unification were mere stepping stones. In fact it is the interdependence of the two fields that shows clearly the other major point of the unification. Waves of light were merely waves of electromagnetic occurrences. The equations imply that periodic changes in the electric or magnetic field travel through space at the speed of light with a changing electric field producing a changing magnetic field, thereby producing a changing electric field, and so on. What is true of light is true of any electromagnetic wave in the spectrum; the magnetic field is perpendicular to the direction of the wave front, the electric field is also perpendicular to the direction of the wave front, and the two fields are perpendicular to each other.[3]

The development of the electromagnetic field concept is widely regarded as the beginning of the breakdown of the old dichotomy of atoms and the void that was central to the seventeenth-century cosmology; for the concept of a field of force means that space is comprised of various stresses and tensions that transmit energy. The notion of empty space as the mere vehicle of spatial interconnections is therefore abandoned as a fundamental principle of physical explanation. The field is rather a medium by which electromagnetic energy has an effect over a distance. It pervades space and contains recognisable routes of energy. James Clerk Maxwell in his *Treatise* attributes the breakthrough to Faraday: 'He never considers bodies as existing with nothing between them but their distance, and acting on one another according to some function of that distance. He conceives all space as a field of force, the lines of force being in general curved, and those due to any body extending from it on all sides, their directions being modified by the presence of other bodies' (1873: 177). In place of the vacuum, Maxwell viewed the undulations as occurring in an ethereal substance, and not of the gross matter ([1865] 1982: 34). Faraday's emphasis on a 'medium' as the embodiment of electromagnetic forces is expressed in Maxwell's theory as the ether, a sea of space through which the transmissions of light, heat and radio waves are possible.

So, with the introduction of the field concept our picture of reality ceased to be purely atomistic, but the old idea of material points stubbornly remained. It is clear that Maxwell's theory required two kinds of entities: the electromagnetic field, spread continuously through space, and varying smoothly in space and time, and highly localised points of matter that both create and are acted upon by the field. While it might be disputed whether the concept of the field by itself necessitates an event ontology, Whitehead contends that the concept of the energy field introduced a radically new idea in physics that points to a network of events as the underlying structure. This combined with Einstein's theory of relativity and quantum theory sealed the fate of the substance ontology.

EINSTEIN-MINKOWSKI SPACE-TIME

Building on the achievements of Faraday, Maxwell and Lorentz, the next great leap of unification in modern physics came with Einstein's conception of physical reality as a continuous electromagnetic field in four-dimensional space-time. As he says: 'The special theory of relativity, which was simply a systematic development of the electro-dynamics of Clerk Maxwell and Lorentz, pointed beyond itself . . .' (1956: 57). While it is clear that the special theory of relativity was Einstein's result of working out the implications of Maxwell's equations such that they take the same form in any coordinate system, his general theory of relativity was his crowning achievement for the development of a theory of gravity. Einstein dismissed the ether as unnecessary, postulated the absence of any absolute frame of rest, and established the invariance of Maxwell's equations in all inertial frames of reference.

Einstein's special theory of relativity began with two basic postulates that were central in his attempt to reconcile Newton's laws and the laws describing electromagnetic phenomena: (1) the equivalence of all inertial reference frames, and (2) the constancy of the speed of light. Einstein resolved the apparent conflict between these two postulates by adjusting the measurement of time and space so that the same flash of light has the same velocity in all inertial reference frames however they are travelling with respect to one another ([1920] 1962: 31).[4] Clocks slow down and measuring rods contract with relative motion, but the velocity of light is the same across the reference frames.

The constancy of the velocity of light, commonly denoted by c, was the one indisputable fact that Einstein drew from the failure to discover any motion of the earth relative to the 'light medium,' that is, the ether.[5] If the velocity of light is constant regardless of the motion of the earth, then it must also be constant irrespective of the motion of any planet, star or galaxy in the universe. Einstein generalised this insight in his assertion that the laws of nature are the same for all uniformly moving systems, including mechanical laws and the laws governing light and other electromagnetic phenomena.

Given a set of inertial frames all moving with constant velocity with respect to one another, events that are simultaneous in one inertial reference frame cannot be simultaneous with respect to any other inertial frame moving with respect to this one (ibid.: 26). This entails that what is 'now' or 'present' must be frame-dependent. 'Now' differs from one inertial frame to another. In addition, Einstein insisted that no inertial frames are to be treated as privileged or special but instead all are to be regarded as equivalent. Since 'now' in our inertial frame of reference is no more privileged than the 'now' in another inertial frame of reference travelling at some uniform velocity with respect to ours, there is no way to divide off an absolute past from an absolute future. In other words, it is impossible to make instantaneous cuts perpendicular to the time axis because doing so would result in different instantaneous spaces in different inertial systems, none of which constitutes a privileged present. The ultimate result of the denial of the simultaneity of distant events means there is no unique 'Everywhere-Now'. Newton's three-dimensional absolute space and absolute time are impossible. The sinking of the Titanic in 1912 happens now for an extraterrestrial man on a distant planet in a distant galaxy in motion, with respect to an earth man now in 2014. For him, the earth man in 2014 is well over a hundred years in his future. So, as Hermann Minkowski saw it, the only way to make sense of this consequence of Einstein's theory is to view the universe spread out in time as well as spread out in space. In this four dimensional space-time, 'now' is on a par with 'here'. As Einstein makes the point:

> Now before the advent of the theory of relativity it had been assumed in physics that the statement of time had an absolute significance, i.e., that it is independent of the state of motion of the body of reference. But . . . this assumption is incompatible with the most natural definition of simultaneity; if we discard this assumption, then the conflict between the law of propagation of light *in vacuo* and the principle of relativity disappears. (ibid.: 27; also see 30)

In this manner, Einstein says, the achievement of the special theory of relativity can be characterised thus:

> It has shown generally the role which the universal constant c (velocity of light) plays in the laws of nature and has demonstrated that there exists a close connection between the form in which time on the one hand and the spatial co-ordinates on the other hand enter into the laws of nature. (1956: 45)

Einstein eliminated Newton's absolute time because there was no longer any need for an objective, independent flow of time and no basis for time independent of frames of reference. Space-time is required by the special theory because while spatial and temporal quantities vary according to a frame of reference, there is an invariant quantity – the space-time interval – that can be dissected into spatial and temporal intervals in innumerable ways according to the frame of reference of a given observer. Although the different observers who are travelling along different paths through space and time differ on their observations about space and time, they concur on the paths described in the four-dimensional space-time.

As noted above, the unification of space and time into space-time is due to the work of Minkowski. He formulated the mathematics of space-time and introduced the idea of the light cone as a way of visually representing space-time from the perspective of a particular observer. With the shift from absolute time, which is composed of instants, to the relativistic light-cone structure, every event is conceived as having a light cone that divides the time-like and light-like directions into two classes: the future and the past light cones. Once events are understood to be the basic elements of the spatio-temporal ontology, the next step is to specify the geometry of space-time as the physical structure of these events.

The geometry of special relativity, Minkowski space-time, can be specified by reference to some special coordinate system, Lorentz coordinates. The manner in which the Lorentz coordinates relate to the geometry of Minkowski space-time roughly corresponds to the way that Cartesian coordinates relate to Euclidean space. Minkowski himself explicitly acknowledged that he arrived at his space-time concept through a purely mathematical line of thought. He began with the idea of combining a spatial point located at the coordinates x, y, z, with a point in time t, to define a *world-point*. He then defined the multiplicity of all such points as defining the *world*. In this manner, at everywhere and everywhen there is something perceptible and no 'yawning void anywhere'. With variations dx, dy, dz of the space coordinates of this point corresponding to a time element dt, we obtain, he says, an image of the everlasting career of the substantial point, a curve in the world, or a *world-line*, of which the whole universe is conceived to be the multiplicity of world-lines. The physical laws find their expression as reciprocal relations between these world-lines (1923: 75–6).

In reference to four-dimensional space-time, Einstein noted that time 'is robbed of its independence' (1962: 56). The time coordinate plays exactly the same role as the space coordinates. And in one of Minkowski's oft-quoted remarks, he famously predicted that: 'space by itself, and time by itself, are doomed to fade away into mere shadows, and only a kind of union of the two will preserve an independent reality' (1923: 75). The crucial point to note is how time has become geometricised in Minkowski's synthesis of space and time.

Everything has been changed in Einstein's theory to accommodate the strange behaviour of light and the various puzzles relating to clocks and measuring rods in the separate frames of reference, each of which constitutes a separate world-line in the geometry of Minkowski space-time. Clocks, namely some observable physical device by means of which numbers can be assigned to events on the device's world-line, are objects spread out in the four-dimensional manifold. There is no longer any sense to the idea of clocks measuring an objective, absolute time. Instead the universe is comprised of an infinite number of world-lines of which the clocks behave more like odometers on cars, each running at a different rate. The same holds for measuring rods since the space measured in one frame will differ from another. The only thing that remains constant once again is the speed of light.

Einstein's special theory applies to all physical phenomena except gravity. The general theory that Einstein said pointed beyond special relativity provides a different conception of the law of gravitation. It was here that Einstein developed the field equations that describe the fundamental interaction of gravitation as a result of a curved space-time. General relativity generalises special relativity and Newton's law of gravitation in order to provide a unified description of gravity as a geometrical property of space-time. The curvature of space-time is directly linked to the energy and momentum of matter.

As mentioned above, what Einstein regards as a disturbing dualism of material points and fields was necessitated in his view by the gravitational field. He writes:

> According to the general theory of relativity the metrical character (curvature) of the four-dimensional space-time continuum is defined at every point by the matter at that point and the state of that matter. Therefore, on account of the lack of uniformity in the distribution of matter, the metrical structure of this continuum must necessarily be extremely complicated. (1923: 183)

While this passage appears to indicate that Einstein held some version of a material ontology, one must keep in mind that objects in space-time, whatever their mass happens to be, are stretched in four-dimensions. They no longer endure as the selfsame objects thereby admitting, as Aristotle put it, contrary qualifications. Time is now internal to an object instead of being external and independent of it.

In his Appendix V of *Relativity: The Special and the General Theory*, written later in 1952, Einstein said that, in accordance with Descartes' concept of the plenum, he wished to show that space-time is not something that has a separate existence, independent of the actual objects of physical reality. 'Physical objects are not *in space*, but these objects are *spatially extended*' (1962: vi). And in Appendix V, in connection with our experiences and the events to which they correspond, he writes: 'It is just the sum total of all events that we mean when we speak of the "real external world"' (ibid.: 140). Similarly when he discusses Minkowski's four-dimensional space-time, he described it as being 'composed of individual events' (ibid.: 55). Given the implications of special and general relativity, it is clear that the concept of material substance had to be abandoned along with absolute space, time and simultaneity. And as Einstein struggled with his unified field theory, he sought to show that particles could be understood as intense regions of the field.

The space-time fusion coupled with the annihilation of the universal present or cosmic now entails a fundamental commitment to an ontology of events. Nicholas Maxwell notes that Einstein formulated special relativity originally in 1905 in terms of an object ontology because the reference frames are characterised in terms of persisting objects, that is, rods and clocks, but then when Minkowski reformulated special relativity along the lines of the geometry of space-time, Einstein eventually accepted the space-time view as an essential step to formulating general relativity. With respect to what he calls 'objectism', or the substance metaphysics, Nicholas Maxwell writes:

> At any instant, here and now, there must be a cosmic-wide state of the universe, indeed *the universe* at that instant. But this clashes with [special relativity]. Observers (or objects) in relative motion here and now have, associated with them, different cosmic-wide presents or 'nows'. In denying that there is any such thing as a privileged reference frame, [special relativity] denies that there is, associated with any space-time point or event, a privileged, cosmic-wide instantaneous 'now' which divides all other events into those that are 'past' and 'future'. Any space-like hyperplane that passes through the point or event in question is as good an instantaneous 'now' as any other. Thus [special relativity], if true, rules out objectism, and demands that eventism must be accepted – a version of eventism that holds that space-time is Minkowskian. (2006: 233)

While general relativity implies that special relativity is false since the latter asserts that space-time is flat and the former asserts that space-time is curved, Einstein says that general relativity presumes the validity of special relativity 'as a limiting case and is its consistent continuation' (1956: 42). Given the two points made above, namely the fusion of space-time and the rejection of a universal or cosmic present, general relativity must likewise be understood as affirming the view that events are basic.

Others have affirmed that an ontology of events is the obvious consequence of the theory of relativity. Milič Čapek, for example, says:

> . . . the replacement of the term 'particle' by 'event' is justified not only by
> the obvious empirical inadequacy of the former, but also because it appears
> logically implied by . . . the concept of space and time. If space and time are
> fused together into the dynamic unity of time-space, which itself was fused
> with its concrete physical content, i.e., matter and energy; if furthermore,
> there is substantial evidence for the pulsational character of time-space, the
> character which matter itself in virtue of its fusion with time-space must
> share, the assertion that what we used to call 'particle' is in truth a string of
> successive events will become less paradoxical. (1961: 259)

For Čapek, those who continue to affirm the substance-like character
of particles have failed to realise the truly revolutionary nature of the
paradigm change in modern physics.[6]

When Whitehead took up the ontological implications of the theory
of relativity, he attributed to Minkowski's genius the full generality of
'the conception of a four-dimensional world embracing space and time,
in which the ultimate elements, or points, are the infinitesimal occurrences in the life of each particle' (1948: 306). These ultimate elements
are events, according to Whitehead, and he made it clear that we must
not conceive of events as existing in space and time but rather think
of events as the units from which space and time are abstracted. As
he sought to reconstruct the Einstein-Minkowski space-time, he began
with event-particles, which are events with all their dimensions ideally
restricted (1922: 29). What distinguishes Whitehead's view from the
classical view is the replacement of one unique serial order of time with
an indefinite number of times as established by the frames of reference
in the four-dimensional manifold (1920: 178). So, while Whitehead
was critical of Einstein's operational procedures involving the transmission of light signals in arriving at the metrical character of the
space-time continuum, he did absorb the most essential aspect of the
Einstein-Minkowski space-time in formulating the event ontology.[7] As
we shall see in the next chapter, however, Whitehead's event theory
underwent two phases of development, one in which extension in the
four-dimensional space-time is basic and another in which temporal
becoming is basic.

In summary then, there are three related aspects of relativity theory
that imply an ontology of events rather than substances. First, the denial
of absolute simultaneity implies that there can be no privileged cosmic
now running along the instants of absolute time, but instead there must
be a multiplicity of frames of reference within space-time. Second,
the four-dimensional space-time has point-events as its foundation.

Indeed, these point-events constitute the reference points for the light cones and serve as the basis from which the geometry is constructed. Third, any material object within space-time must be conceived as an entity stretched out in four dimensions with both spatial and temporal parts. As a consequence of this, time can no longer be treated as being external to and independent of a material object; rather time is now an embedded feature of each object. It is, quite simply put, a sequence of events.

THE EARLY QUANTUM THEORY

By 'early quantum theory', I have in mind the atomic models and theories of Planck and Bohr as opposed to the new quantum theory advanced by De Broglie, Heisenberg, Schrödinger, Born, Dirac and others beginning in the late 1920s. Since the event ontology was, for the most part, influenced by the early theory, I focus here on that development and its influence specifically on Whitehead's metaphysics. Whitehead's early event theory is based on Maxwell's electromagnetic theory and Einstein's relativity theory, but his later process metaphysics is more reflective of the dynamic-field ontology of quantum physics.

As noted in Chapter 1, quantum theory lent further support to the development of event ontologies due to the un-thing-like behaviour of elementary particles such as electrons and protons. The central mystery of quantum theory is indisputably ontological in character due to the fact that particles appear to be located at a point in space in a particle-like fashion *and* spread out over space in a wave-like fashion. But how can they be both? Particles have been conceived traditionally as solid substances such that two particles hitting a detector at the same point would result in double the mark of one particle, whereas waves impacting at the same place might cancel one another out in the phenomenon called 'interference'. The double-slit experiment, which Richard Feynman said presents 'a phenomenon which is impossible, *absolutely* impossible, to explain in any classical way', is key to understanding quantum mechanics (1964, III: 37–1). An experimental set-up where a stream of electrons emanates from a source and is directed toward two slits reveals at the detector on the other side that the impacts of the electrons cancel one another out in a wave-like interference pattern. This kind of experimental result might be expected to be produced by light but not particles which, according to classical mechanics, would behave more like bullets fired from a gun. So, this raises the question about the ontology of this wave/particle duality.

For Whitehead's metaphysics, it was clear that the very concept of energy quanta meant more trouble for the substance theory and for its counterpart in physics, that is, materialism. As he says:

> [M]etaphysical concepts, which had their origin in a mistake about the stone, were now applied to the individual molecules. Each atom was still a stuff which retained its self-identify and its essential attributes in any portion of time – however short, and however long – provided that it did not perish. The notion of the undifferentiated endurance of substances with essential attributes and with accidental adventure was still applied. This is the root doctrine of materialism: the substance, thus conceived, is the ultimate entity. ([1929] 1978: 78)

Quantum theory, however, revealed that nature was no longer cooperating with this conception. Whitehead writes:

> The mysterious quanta of energy have made their appearance, derived, as it would seem, from the recesses of protons, or of electrons. Still worse for the concept, these quanta seem to dissolve into the vibrations of light. Also the material of the stars seems to be wasting itself in the production of the vibrations. (ibid.: 78–9)[8]

And again:

> . . . the change . . . is the displacement of the notion of a static stuff by the notion of fluent energy. Such energy has its structure of action and flow, and is inconceivable apart from such structure . . . Mathematical physics translates the saying of Heraclitus 'All things flow', into its own language. It then becomes, all things are vectors. (ibid.: 309)

In short, the modern notion of the atom has been 'dematerialised' such that the formal atoms of modern theory are no longer construed to be self-identical substances across time.[9] At the heart of the matter is the notion that energy of all types occur in quanta or minimal packets. Atoms are understood in terms of waves of radiation that they can emit or absorb and both processes of emission and absorption occur within non-uniform spans of time. The reason why a simple particle theory of the atom, using ordinary mechanics and electromagnetic theory, is no longer feasible is that the electron cannot be considered simply as a particle. Since it cannot have both a well-defined position and velocity it must be seen in part as a wave. The orbits of electrons are to be regarded as series of detached positions rather than continuous lines. Indeed, it is from this modification that the locutions 'quantum leaps' or 'quantum jumps' were derived. The idea of classical mechanics that particles have a definite size, shape and position was therefore abandoned. The emphasis now is placed upon pulses of energy that have an approximate location in space-time and interact in fields that bear and transmit the forces of nature.

Because of the extremely short lifespan of the subatomic particles of modern physics as compared to the constancy, indestructibility and eternality of the classical conception of atoms, Čapek again argued that events replace particles. What is still called 'particles', he argues, have neither particle nor corpuscular properties. The use of accelerators in the search for the ultimate building blocks of nature, the true atoms, has revealed higher and higher energies of subatomic particles.[10] The lifetime of most of these particles is so exceedingly brief, their duration being almost equivalent to a flash of light, that these quasi-instantaneous entities are better understood as events (1961: 258–9, 285). Mesons, for example, 'last' only 10^{-16} of a second. And according to the latest observations, the Higgs boson decays into other particles almost immediately. Our investigations into the nature of elementary particles reveal properties and only properties existing in ever-smaller units of space-time.

The emphasis on pulses of energy in quantum theory suggested to Whitehead that: 'Matter has been identified with energy, and energy with sheer activity' ([1938] 1968: 137).[11] He identified the central mistake of classical physics as the 'fallacy of simple location'. This is the supposition that a material entity is simply located, that is 'in a definite finite region of space, and throughout a definite finite duration of time, apart from any essential reference of the relations of that bit of matter to other regions of space and to other durations of time' (1925: 72). As noted above in connection with Newtonian mechanics, this view is sometimes called 'the doctrine of external relations'. Whitehead says, however: 'in the modern concept the group of agitations which we term matter is fused into the environment. There is no possibility of a detached, self-contained local existence' ([1938] 1968: 138). He therefore avoids simple location with his notion that nature at its most basic level is constituted of epochal becomings that create the space-time continuum. In other words, the basic events, which Whitehead calls 'actual occasions', are atomic quanta of becoming that cannot be located in a definite place in space and time, but rather occur as durations of space-time that come all at once and are connected to antecedent events via their 'prehensions' of one another.

In accordance with the early quantum theory, Whitehead maintained that all physical experience happens in leaps or definite epochs of becoming. What is particularly important about the findings of quantum theory is that, at the very base of nature, the discontinuous existence of microscopic particles forms the continuous existence of macroscopic bodies. Undoubtedly, Whitehead found this idea crucial in explaining how his basic ontology of events accounts for various levels of enduring objects. He explicitly says in fact that his cosmological theory is 'perfectly consistent with the demands for discontinuity which have been urged from the side of physics' (1925: 171). Physical reality is, at most, quasi-continuous, as successive leaps or vibrations of energy fuse together to form physical objects perceived by us as continuous.

In comparing the characteristics of quantum phenomena with the events that Whitehead proposed as the fundamental units of ontology, it should be clear that he is not equating the physical activity of subatomic particles with the basic events. What is observable to the physicist is rather multiple interactions of events with an electromagnetic character. As he made this point, he writes: 'The notion of physical energy, which is at the base of physics, must then be conceived as an abstraction from the complex energy, emotional and purposeful, inherent in the subjective form of the final synthesis in which each occasion completes itself' (1933: 239).

According to Bohr's interpretation of quantum phenomena, both the wave and particle conceptions are used in a complementary manner depending on the experimental situation but without any speculation about the underlying picture of nature that is producing the phenomena. Whitehead, however, provides an ontological interpretation of quantum entities with his epochal theory of becoming. Nature is comprised of atomic, unit events that are more wave-like, but the objects revealed in quantum systems, such as the double-slit experiment, have characteristics that are both wave-like and particle-like. The wave concept is more fundamental in the sense that the quantum entity is not an object that is simply located but rather occurs as a slab or duration of space and time. What is observable in experiments is understood to be a manifestation of characteristics belonging to these underlying events.

THE STATE OF A UNIFIED THEORY

In summing up the state of theoretical physics in 1948, Einstein wrote: 'the growth of our factual knowledge, together with the striving for a unified theoretical conception comprising all empirical data, has led to the present situation which is characterized – notwithstanding all successes – by an uncertainty concerning the choice of the basic theoretical concepts' (1948: xix). The question of whether quantum theory or general relativity is the more basic theoretical concept remains unanswered and continues to be a central issue of debate. As quantum theory continued to develop from its rudimentary beginnings, it became increasingly successful in its empirical predictions, and up to the present it has remained one of the most secure theories in physics. But it has also been mired in baffling mysteries regarding the fundamental conception of nature that it is meant to explain. Even more to the point, according to Einstein, it has remained the main stumbling block to any successful unification.

When Einstein sought his unified field theory in the 1920s the weak and strong nuclear forces had not yet been discovered. Thus his project was the simpler one of unifying general relativity with electromagnetism. But he also believed that the unification of gravity with electromagnetic phenomena was linked to resolving the paradoxes of quantum mechanics because a complete theory required consistency throughout the theoretical framework. Yet as Einstein continued to explore solutions, rejecting one hypothesis after another – all supposing the fundamental correctness of general relativity – he ignored developments in quantum mechanics and distanced himself from the mainstream of physics. He expressed this view in the final paragraph of his *Relativity: The Special and the General Theory* when he wrote: 'one should not desist from pursuing to the end the path of the relativistic field theory' as a means of resolving fundamental difficulties toward the goal of unification with quantum theory. For in his view, the consensus of opinion regarding the statistical nature of quantum theory together with the corpuscular-wave duality constituted a 'weakening of the concept of reality' (1962: 157).

While theoretical physics has had limited success with unifying the non-gravitational forces that operate on a subatomic scale,[12] Einstein's basic problem remains unsolved in that the forces described in the partial theories of the twentieth century have continued to resist all attempts to unify them. Many physicists today agree that the current contenders for the final unification – Gauge, String and M theories – are exotic mathematics with no connection to physics, so we are no closer to a solution to the Big Problem. The quantum theory of gravity, which attempts to explain the force of gravity by principles of quantum field theory, has been another contender for a Theory of Everything, but it too remains highly speculative. In the terms of Thomas Kuhn's philosophy of science, we have not achieved a comprehensive paradigm unifying the whole of physics but rather have sub-paradigms that unify different aspects of physics (1962). Whitehead and Russell thought that an ontology of events would provide a general unifying concept but this was early in the twentieth century and the fine-grained, micro-structured, puzzle-solving that Kuhn characterised as necessary to paradigm construction had not yet come into focus. There are, however, several physicists who have found in Whitehead's work a direction for the micro-structure. These we shall take up in Chapter 6 below.

4. *The Revisionary Theory of Events*

This is often the way it is in physics – our mistake is not that we take our theories too seriously, but that we do not take them seriously enough.
(Steven Weinberg)

When Whitehead, Russell, Broad and advanced novel formulations of an event ontology, it was clear that they all believed science had outgrown the substance theory of Aristotle and his followers. As Russell points out, belief in substance seemed warranted as long as physics assumed one cosmic time and one cosmic space, but this view was radically altered with the arrival of the Einstein-Minkowski concept of space-time ([1927] 1934: 286).

In this chapter I examine the affinities and contrasts in the event theories advanced by Whitehead, Russell and Quine, all of which originate from the revolution in twentieth-century physics. The revisionary theory of events overthrows the descriptive theory, according to which events are dependent on substances. Events, under this new theory, are basic, and substance, as an ontological category, is eliminated.

WHITEHEAD'S EARLY ONTOLOGY OF EVENTS

Whitehead first introduced his theory of events in his *Principles of Natural Knowledge* in response to Maxwell and Einstein. He writes: 'Modern speculative physics with its revolutionary theories concerning the natures of matter and of electricity has made urgent the question, What are the ultimate data of science?' (1919: v). This enquiry takes the form of a classification of natural entities that are posited for knowledge in sense awareness (1920: 49). The thesis he advances for the unification of scientific knowledge is that the ultimate facts of nature, in terms of which all physical and biological explanation must be expressed, are events connected by their spatio-temporal relations, and that these relations are in the main reducible to the property of events that they can contain (or extend over) other events which are parts of them. (1919: 4)

'The whole object of these lectures', he writes in *The Concept of Nature*, 'is to enforce the doctrine that space and time spring from a common root, and that the ultimate fact of experience is a space-time fact' (1920: 132). In *The Principle of Relativity*, Whitehead called this attempt to unify the natural sciences under one concept 'pan-physics' (1922: 4).

He defined an event as 'the specific character of place through a period of time' (1920: 52). What we sense in the passage of nature is both spatially and temporally thick and is understood as a relatum in a complex structure of events that forms the extended universe. Broad similarly defines an event as 'anything that endures at all, no matter how long it lasts or whether it be qualitatively alike or qualitatively different at adjacent stages of its history' (1923: 54).

With regard to the foundations of dynamical physics, Whitehead develops the theory of events via the contrast between Newton and Maxwell. Maxwell retained the idea of an all-pervading ether in order to account for gravitational, electrostatic and magnetic attractions. His equations of the electromagnetic field, Whitehead says, presuppose events and physical properties of apparently empty space, which necessitates the 'metaphysical craving' of the ether. There must be something in the seemingly empty space to which these properties belong. Moreover, he thinks that the same ether is required by the apparently diverse optical and electromagnetic phenomena (1919: 20–1). The ultimate facts of Maxwell's equations are the occurrences of volume-densities and charged velocities located at the space-time points within the neighbourhood surrounding a given space-time point. But this, Whitehead writes, is just to say that the ultimate facts contemplated by Maxwell's theory are events occurring throughout all of space (ibid.: 24–5). Instead of a material ether, there is an 'ether' of events, which is just the continuity of nature expressed as a network of events. The upshot, in contrast to Newton, is that we must not think of events as changes in material at a time and place but instead should construe space, time and material as arising out of the uniform structure of events.

The revolution resulting from Maxwell and Einstein's work provided Whitehead with the opportunity to rethink the foundations of scientific knowledge. The time was ripe to begin afresh with a theory that would unify the entire spectrum of physical phenomena from the microscopic forces at the subatomic level to the macroscopic, large-scale structure that is determined by gravity. To accomplish this, Whitehead began with his method of identifying the entities of sense awareness and then arriving at the precise concepts of mathematical physics and geometry via the logical construction he called 'extensive abstraction'. His aim was to fill the gulf between perception and physics, between the rough world of percepts and the smooth world of mathematical physics.[1] The abstractions of mathematical physics, however, are not directly perceived; they are rather derived by the logical process of progressive refinement of events to arrive at ideal entities like points, lines and planes that function as the elements of the geometry of space-time.[2]

In opposition to the atomistic framework of British empiricism, Whitehead reshapes the notion of experience so as to accord with the phenomenon of the specious present – the new or radical empiricism pioneered by William James and the Gestalt psychologists. Nothing exists at an instant. Instead of beginning with points in space or material configurations at an instant and then building our picture of the physical universe outward, the passage of nature and the experience of an observer are placed on a closer footing. Both occur as temporal slabs. For Whitehead, the specious present is an example of the unity of an event that then becomes the basis for understanding the unity of nature. This is not to say that physics is confined to the events of the specious present, but rather that the idea of spatio-temporal spans, of whatever magnitude, must replace points and instants, if indeed our science of nature is to be based on observation.

Within the duration of one's specious present, one discerns subordinate events by whole-part relations, and by working outward from the specious present, the larger space-time structure becomes known only as *relata* in relation to the entities in the discerned field. This is what Whitehead calls the 'discernible'. For example, the orbits of planets in distant galaxies are not at present discerned but are, in principle, discernible. The discerned is always part of a broader field of that which is discernible. The notion that events are interrelated to form the space-time structure of nature is what he calls his 'doctrine of significance' (1920: 51). Space-time is an abstraction from the concrete order of events (1922: 21).

Whitehead's ontology and epistemology are based on the distinction between repeatable and non-repeatable entities, both of which are known in experience. Non-repeatable entities are events and are, for Whitehead, the particulars of space-time. They can only happen once. Repeatable entities he calls 'objects', and these are discriminated into 'sense-objects,' such as individual colours, sounds or textures, 'perceptual objects' such as the ordinary macroscopic bodies of perceptual experience, and 'scientific objects' such as electrons and molecules. Objects are the *recognita* amid events (1919: 81). They are the things in nature that can be again (1920: 144). For example, the value and intensity of a particular instance of white, the physical body with which I write on the blackboard and the molecular structure of calcium carbonate ($CaCO_3$) all continue to characterise a sequence of events that we recognise as a piece of chalk.

In accordance with Whitehead's proposed theory, anything that appears to exhibit permanence and an abiding structure in nature must be explained in terms of event processes. Everything in the universe, from medium-size dry goods to planets and galaxies, is reinterpreted as patterns of properties that are repeated in event sequences. 'Things', as we ordinarily understand them, are postulated by Whitehead to be relatively monotonous patterns in events. They are bundles of energy that more or less maintain their characteristics and consequently form 'space-time worms' (or, following Minkowski, 'world-lines') in the four-dimensional manifold. As Broad put the point: 'A thing . . . is simply a long event, throughout the course of which there is either qualitative similarity or continuous qualitative change, together with a characteristic spatio-temporal unity' (1923: 393). Whitehead's example from *The Concept of Nature* is the Ancient Egyptian obelisk, 'Cleopatra's Needle', on the Embankment in London. This enormous and seemingly permanent structure may not appear to be an event, especially when compared to a short duration of a traffic accident below. He argues, however, that its abiding structure is simply a relatively stable situation in the stream of events constituting this permanence of character and so the difference between it and the traffic accident is merely one of time span (1920: 165–7).

From the above it should be clear that Whitehead's event theory reverses the ontological priority of the common-sense conceptual scheme. Events support objects rather than the other way around. They are the basic, concrete entities upon which we perceive the continuity of objects across time. Strawson's idea of identification and re-identification is retained as a linguistic convenience; however, it is not substance but patterns of properties that are repeated in diverse events that we recognise and it is these patterns that are the primary objects of reference. In fact, for Whitehead, science (or for that matter, any type of communication) would not be possible without the objects that give events structure and definition. Laws of nature are in effect precise descriptions of patterns of objects in the passage of events.

In Whitehead's ontology, events have no independent existence. They blend into one another with the passage of nature (1919: 73–4). In this respect, nature forms an interdependent system of internally-related events that are ordered by the primary relation of 'extending over'. At each moment, nature is an all-comprehensive event or 'fact' within which we discriminate constituent events and objects as 'factors' (1920: 13). The fact that sense awareness does not apprehend definite spatio-temporal limits in events, however, does not imply that nature is an undifferentiated whole. Whether events are longer or shorter, extended over or extending over, will depend on how they are described, but the properties or 'objects' ingredient in events provide the natural boundaries (ibid.: 144, 172–3). Without the description, the demarcation of events would be arbitrary. So, for example, 'the French Revolution' is a description of a long, vastly complicated event that extends over the storming of the Bastille, and the storming of the Bastille extends over the release of the prisoners. But the same spatio-temporal region could be described in innumerable ways by different historians, as well as by the political scientists, sociologists and economists who have provided radically different descriptions as a consequence of their having different purposes. As long as the primary relation of extending over is used to mark off a certain amount of qualitative similarity in the events described, there is no limit to how the space-time regions might be individuated.

So, it is quite clear that Whitehead's theory of events is an ontological generalisation of the energy field. Just as particles are not independent entities but rather parts of the field that act on and create the field, objects are not conceived as independent entities; they are rather the ingredients that give events their structure. Just as the field is spread continuously and smoothly throughout all of space and time, the whole system of events is similarly uniform and continuous with no clear breaks nor any beginning nor end.

Returning to the idea that the entities postulated by Maxwell's theory are fields and material points, one can see that Whitehead is giving an ontological interpretation of these entities in terms of his events and objects. The space-time structure of the electromagnetic field is understood as being a uniform continuity of events. The lines of force that mark the stresses and the vector paths in the field are construed to be objects ingredient in events, that is, as physical properties of the field that provide for quantitative interpretation. 'Scientific objects', such as charged particles, are not understood as material points on a par with the field; they are re-classified within the general category of properties of the field. It is clear that such scientific objects are not actually observed in sense awareness, but their inferred existence is required as a part of scientific theory, as a refinement of sense awareness. What is observed is the character exhibited in events. As Whitehead puts it:

> The electron is its whole field of force. Namely the electron is the systematic way in which all events are modified as the expression of its ingression. The situation of an electron in any small duration may be defined as that event which has the quantitative character which is the charge of the electron. (1920: 159)

So, particles such as electrons are not to be thought of on the model of individual substances with properties, as entities with charge; rather they are rhythmic repetitions of charge in events.

Whitehead rejected matter as a fundamental ontological category because he argued that it did not stand up to empirical examination. This was also the main basis of his disagreement with Einstein's theory that space-time is heterogeneous due to the peculiarities in the distribution of matter throughout the universe. Whitehead instead contended that the metric of space is defined in terms of objects ingredient in events such that no essential connection with the apparent distribution of matter is required. The objects exhibited in events form patterns and the space-time of our perceptions is conceived as continuously uniform with the more refined space-time of scientific objects.

In this middle period of Whitehead's work, his pan-physics, it is clear that he has accepted the Einstein-Minkowski view of a four-dimensional space-time, but, as noted above, with some reservations about the empirical foundations of the theory. In fact, when he developed his own theory of relativity, he said with reference to Einstein and Minkowski et al. that 'the worst homage we can pay to genius is to accept uncritically formulations of truths which we owe to it' (1922: 88). Although Whitehead's construction was largely founded on the ideas of Einstein and Minkowski, he diverged from them in one important respect: he based his own four-dimensional system on durations. Indeed, this is one important application of his method of extensive abstraction. Durations are events with finite temporal and infinite spatial extension (1919: 110). They are parallel when any two of them are extended over by a third; otherwise they are non-parallel. Families of parallel durations and the refinement of these into parallel moments constitute the succession of time in any one time-system. Time is the abstraction of the parallel moments from the sense of passage within parallel durations. And, within space-time as a whole, there is an indefinite number of families of parallel durations constituting different time systems, none having any members in common.

Whitehead now brings into play his concept of 'cogredience' to serve as the basis of motion and rest in his theory of relativity. Cogredience is the extension of a finite event throughout a duration, or as he puts it in *The Concept of Nature*, 'the perseveration of unbroken quality of standpoint within a duration' (1920: 110). If the duration is the content of the specious present of an observer, a cogredient event is the part of this content that lasts through the whole duration and does not change its position relative to the body of the percipient during the specious present. The finite event can be some particular body or event-particle that occupies successive positions in one time-system. A body is at rest for an observer whose specious present includes that body within the permanent space of its time-system. But relative to the space of another time-system, the body is moving in a straight line with uniform velocity (ibid.: 114). Motion and rest depend on the time-system that is fundamental for the observation. This stratification of time-systems corresponds roughly to Einstein's inertial frames of reference within the four-dimensional manifold and marks one of the most important departures from Newtonian physics in which there is only one time-system, that is absolute time. The multiplicity of time-systems in Whitehead's theory also suggests, in accordance with Einstein and Minkowski, that past, present and future are simply relative to some frame of reference and that, within space-time as a whole, all events are ontologically fixed.

WHITEHEAD'S LATER PROCESS ONTOLOGY

As Whitehead continued to develop his event ontology in his later metaphysical works, *Science and the Modern World* and *Process and Reality*, he considerably expanded the scope of his project. Whereas in his pan-physics his aim was to create a unifying concept for the natural sciences, in his later theory he sought a comprehensive scheme that would explain a broader range of phenomena. Whitehead therefore developed a more general conception of events that would explain the evolution of organisms, electromagnetic energy and the higher phases of human consciousness via general metaphysical principles of process and creative advance – a unification of twentieth-century physics, biology and psychology in his philosophy of organism.

'Actual occasions' (or 'actual entities') replace 'events', and 'eternal objects' replace 'objects'. Whitehead's ontology remains dualistic in that he posited two fundamental types of entity: those that are repeatable and those that are non-repeatable.[3] This appears to be the traditional distinction between particulars and universals except that he expressly emphasises in his ontological principle that 'actual occasions are the only *reasons*'; and that everything must be somewhere in actuality ([1929] 1978: 24). It is thus clear that he rejects the transcendental and independent reality of Plato's Forms and agrees with Aristotle's criticism of Plato that the natural world is not merely a representation of the final reality of the perfect, eternal Forms. The most significant change in his system, however, is the emphasis on process. In a crucial note to the second edition of *The Principles of Natural Knowledge*, he says that: 'The book is dominated by the idea . . . that the relation of extension has a unique preeminence and that everything can be got out of it . . . But the true doctrine that "process" is the fundamental idea', was not in his mind 'with sufficient emphasis' at the time that he wrote the first edition. 'Extension', he says, 'is derivative from process, and is required by it' (1919: 202). In his metaphysics, process now becomes basic and the whole structure of the space-time extensive continuum is derivative, the holistic conception of events in his philosophy of natural science is now developed into a pluralistic temporal atomism and the geometrical and primarily spatial thinking of his earlier preoccupation with relativity gives way to a theory that 'takes time seriously'.

The tendency to negate the reality of time is one of the most persistent features of our philosophical and scientific traditions. In this regard, as Whitehead modified his earlier theory, he opposed tradition in the likes of Parmenides, Plato, Spinoza, Einstein and Minkowski by advancing his process metaphysics. 'Taking time seriously' means giving proper emphasis to becoming and perishing in this system. It does not, however, mean that there is a unique serial order in which the universe evolves, for Whitehead retained the most important principles of Einstein's special theory of relativity. With the overthrow of Newtonian physics, absolute simultaneity and absolute time were replaced with the relativistic notion of frames of reference. Whitehead says 'creative advance is not to be construed in the sense of a uniquely serial advance' ([1929] 1978: 35). There is no absolute becoming from a cosmic Now, but rather becoming from the perspective of individual actualities, the actual occasions. There is no absolute past nor an absolute future, but rather past and future relative to each occasion's present becoming.

Whitehead uses the term 'event' in *Process and Reality* to denote 'a nexus of actual occasions, inter-related in some determinate fashion in one extensive quantum' (ibid.: 73) but he makes it clear that an actual occasion is an event of only one member. Everything else – physical objects, space and time – is built up by abstraction from the concrete basis provided by the actual occasions.

Whitehead was well aware that any theory that describes the universe as a process of becoming must address Zeno's paradoxes, specifically the 'Arrow in its flight' and the 'Achilles and the tortoise' paradoxes. If the paradoxes are modified for time and becoming instead of motion as originally conceived, the argument states that any act of becoming can be infinitely divided and thereby results in the conclusion that nothing becomes, that is, time is unreal. Whitehead's 'epochal theory of time' is his rebuttal of Zeno. The temporal quanta must come as wholes – as epochs of becoming – rather than as parts that comprise the units. So Whitehead argues 'in every act of becoming, there is the becoming of something with temporal extension; but that the act itself is not extensive, in the sense that it is divisible into earlier and later acts of becoming which correspond to the extensive divisibility of what has become' (ibid.: 69). In this manner, he defends atomism against the objection of an infinite divisibility in nature. Actual occasions, the basic events, atomise the extensive continuum of the universe, but in order to avoid Zeno's problem we must not conceive of the process of becoming as extensive. Instead, extension must be construed as resulting from the process of becoming.

Process is now explained via the mechanism of 'prehension' or 'feeling'. This concept explains how we get one event from another. The present *subject* absorbs the data of the immediate past or the *objects*, and the many past occasions become a novel one. The emergence of the present from the past is an asymmetrical relation. The past is *internal* or contained in the present but the present is *external* to the past. That is, the present actual occasion is understood to be a process of self-creation by prehending the past, but the past does not prehend the present since the future of that actual occasion does not yet exist.

The difference between symmetrical interdependence and asymmetrical one-way dependence is crucial for obtaining a clear understanding of Whitehead's transition from his early ontology of events to his process ontology. Whereas in the four-dimensional theory events are interlocked into a complex spatial network that underlies the electromagnetic field, once he accepts process as the basic concept the symmetry implied in the interdependence of events is replaced by a fundamental temporal asymmetry. A system of purely internal relations is replaced by one having both internal and external relations and consequently serves as the foundation for the direction of time as distinct from the mere passage of time. Dependence works one way; the present is dependent on the past and not vice versa.

The word that Whitehead uses to describe the process of becoming concrete is 'concrescence', the *growing together* of the many into a novel unity. Within the concrescent process, the actual occasion actively selects (positively prehends) or rejects (negatively prehends) the data presented by the past. He writes: 'The actual entity terminates its becoming in one complex feeling involving a completely determinate bond with every item in the universe, the bond being either a positive or negative prehension. This termination is the "satisfaction" of the actual entity' (ibid.: 44). To be present is to be a subject or a *private* experience that is in the process of becoming until it reaches 'satisfaction', that is, the terminus of its self-creation. Once the subject completes its process of becoming, it becomes an *object* – a determinate, *public* entity that is then available for the prehensions of future actual occasions. Whitehead's use of the term 'object' here refers to what becomes public fact at the outcome of an actual occasion's process of concrescence. An object, in this sense, is what is perceived in the immediate past and is prehended as a unity by a present actual occasion. So, just as the light received from a star is originating from an object that might in fact be long gone, any perception of an object in the immediate present involves the transmission of light from an event that is in the past of the perceiver's experience. Instead of taking years, as in the case of light travelling from a distant star, in ordinary perception it occurs in a fraction of a second. The idea that the perceived object is never a contemporary one can be generalised for all actual occasions in the sense that the genetic process of concrescence involves the prehension of what is completed by the immediate past occasion just before it perishes.

Realism is secured in this system by the objectivity of the past; any perception of nature is a prehension *of* an independently existing, determinate past. Objects, however, are obviously not conceived in the manner in which substances were traditionally conceived to be, but rather as the data of the immediate past that are synthesised into novel unities. For Whitehead, this provides an explanation for the dynamics of the world such as process, change, evolution and, most importantly, an explanation for how novelty occurs. He writes:

> To be actual must mean that all actual things are alike objects, enjoying objective immortality in fashioning creative actions; and that all actual things are subjects, each prehending the universe from which it arises. The creative action is the universe always becoming one in a particular unity of self-experience, and thereby adding to the multiplicity which is the universe as many. (ibid.: 56–7)

The fundamental action of prehension describes how the many, which is the universe disjunctively, becomes one, which is the universe conjunctively. Creativity is therefore the ultimate category of Whitehead's 'Categoreal Scheme'. This expresses the essence of reality. He says:

'This Category of the Ultimate replaces Aristotle's category of "primary substance"' (ibid.: 21). Instead of taking substance as the most basic category of reality, Whitehead replaces this with the activity of creativity as the master principle of becoming.

In order to give his system the broad explanatory power required for a general metaphysics, Whitehead has revised his earlier concept of events such that the events now have internal characteristics such as creativity, subjectivity and selectivity in addition to the external spatio-temporal, mereological and electromagnetic properties. These internal characteristics are interconnected in the sense that the occasion's aim for accomplishing its satisfaction, its creative selection of the data compatible with this aim and its subjectivity (understood as present immediacy) explains what it means to be in a process of becoming determinate. The potential becomes actual as the atomic quanta, the actual occasions, produce novelty from the data of the past. What an artist does in drawing upon past tradition and modifying it to achieve a new expression, form or idea in producing a new painting, sculpture or musical composition is generalised in Whitehead's metaphysics to express the basic activity of all of nature.[4]

The very language that Whitehead designed to express his view of process and creative advance has contributed greatly to the difficulty in understanding him. This, of course, is one of the perils of departing from ordinary language. If, however, one sees this part of his metaphysics as a generalisation of genetics and evolution, it begins to become clear that he is explaining how novelty arises from the multiplicity of ways that the data of the past are synthesised by the present. In contrast to orthodox biology, where evolution by natural selection is explained mechanically or as a matter of blind chance, Whitehead explains change and evolution at a micro-level via a teleological process of selection.

One of the crucial roles of Whitehead's theory of extension is to provide an explanation of how the microscopic actual occasions form the base of the enduring things of perceptual experience – rocks, plants, animals, planets, stars, galaxies and the whole structure of space-time. The main problem is to explain why we perceive the individuals that are typically identified as 'substances' in Aristotle's ontology rather than multitudes of actual occasions. Whitehead makes the transition from the microscopic to the macroscopic world via his notion of transmutation, which occurs when the multiplicity of occasions in any one physical body are prehended as a unity. When we perceive any macroscopic entity, he explains, we prehend an aggregate of many occasions as one final unity. An individual is discerned in the mass of actual occasions present to consciousness by the way the perceiver integrates the many into one transmuted feeling. This is possible because the members of an aggregate share a dominance of characteristics. Whitehead calls such an aggregate of occasions a 'nexus'. When that aggregate nexus forms the dense regions of space, we get what he calls a 'society'.

Whitehead's concept of society is central to his theory of how actual occasions are grouped together. A society is defined as a nexus of social order in which the constituent actual occasions must positively prehend those eternal objects that not only define that specific society but also ensure its continued survival. A society is not simply an aggregate of mutually contemporary occasions but contains multiple lines of inheritance. Societies, as we shall see in the next chapter, can be either simple or vastly complex. The simplest ones are those with 'personal order', in which the members are ordered serially. In more complex ones, there are societies within societies within societies. A cell, for example, would be a complex-structured society that harbours the existence of lower, more specialised societies – molecules, atoms, electrons and so on.

In *Process and Reality*, the concept of the field continues to play a foundational role in Whitehead's theory of extension, but the electromagnetic character of reality only occupies a particular region of what he calls 'the extensive continuum'. Once process becomes the fundamental idea of his scheme, it becomes clear that the dominant character of our cosmic epoch is that the actual occasions all reproduce the electromagnetic character of the epoch. But this only applies to that part of reality that is identified as *our* cosmic epoch. Whitehead writes:

> The arbitrary, as it were 'given', elements in the laws of nature warn us that we are in a special cosmic epoch. Here the phrase 'cosmic epoch' is used to mean that widest society of actual entities whose immediate relevance to ourselves is traceable. This epoch is characterized by electronic and protonic actual entities, and by yet more ultimate actual entities which can be dimly discerned in the quanta of energy. Maxwell's equations of the electromagnetic field hold sway by reason of the throngs of electrons and of protons. Also each electron is a society of electronic occasions, and each proton is a society of protonic occasions. These occasions are the reasons for the electromagnetic laws; but their capacity for reproduction, whereby each electron and each proton has a long life, and whereby new electrons and new protons come into being, is itself due to these same laws. But there is disorder in the sense that the laws are not perfectly obeyed, and that the reproduction is mingled with instances of failure. There is accordingly a gradual transition to new types of order, supervening upon a gradual rise into dominance on the part of the present natural laws. (ibid.: 91)

In his ultimate conception of the universe, Whitehead viewed the laws of electromagnetism as contingent on a general character of our particular cosmic epoch. Beyond our cosmic epoch, he speculated that there would be other types of order.

RUSSELL'S ANALYSIS OF MATTER

Russell is once reported to have said that, with regard to his commitment to sense data as his basic certainty, he would go to the stake for their existence (Sprigge 1994: 74). Whether he was equally confident about the status of events as the basic particulars with which science deals is less clear but, like Whitehead, after their collaboration on *Principia Mathematica* (1910–1913), Russell turned his attention from the foundations of mathematics to the foundations of physics and advanced an ontology of events. He says in his *Our Knowledge of the External World* that: 'Philosophy, from the earliest times, has made greater claims, and achieved fewer results, than any other branch of learning' ([1914] 1956: 11). And in the preface of that work, he writes:

> But something different is required if philosophy is to become a science, and to aim at results independent of the tastes and temperament of the philosopher who advocates them . . .
> The central problem by which I have sought to illustrate method is the problem of the relation between the crude data of sense and the space, time, and matter of mathematical physics. I have been made aware of the importance of this problem by my friend and collaborator Dr. Whitehead. (ibid.: v)[5]

Again in *The Analysis of Matter*, he says in connection with the problem of understanding the relationship between the world of physics and the world of perception that: 'This difficulty has led, especially in the works of Dr Whitehead, to a new interpretation of physics, which is to make the world of matter less remote from the world of our experience' ([1927] 1934: 6). The very prestige of physics is its mathematical precision but as it increases the scope and power of its mathematical technique, 'it robs its subject matter of concreteness' (ibid.: 130). So, quite clearly Russell agreed that setting physics on a stronger empirical foundation was critical with the revolution that occurred in physics; for without 'an interpretation of physics which gives due place to perceptions . . . we have no right to appeal to the empirical evidence' (ibid.: 7).

At the outset of *The Analysis of Matter*, Russell describes the state of physics in 1927 and laments the fact that quantum physics had not reached the stage of deductive rigour that belongs to the general theory of relativity, but nonetheless he expresses the hope that a unified treatment of the whole of physics would soon be possible. The epistemological question that concerns Russell there is a reiteration of the question that Whitehead had previously raised about the ultimate data of science: What facts and entities do we know that are relevant to physics and may serve as its empirical foundation? A philosophical enquiry into physics must begin with our immediate, direct perceptions. And, as Russell emphasised, what we actually perceive is an *event*, that is a volume of space-time that is small in all four dimensions and what we infer from these events are further groups of events. He therefore argued that the metaphysical status of physics would be vastly improved by a theory of the physical world that makes its events continuous with perception and that perception itself is a perception of events (ibid.: 6, 275). Substance, if it exists at all, 'belongs to the realm of mere abstract possibility' (ibid.: 244).[6]

Science, he concludes, is concerned with groups of events rather than with things because 'the objects that are mathematically primitive in physics, such as electrons, protons, and points in space-time, are all logically complex structures composed of entities which are metaphysically more primitive, which may be conveniently called "events"' (ibid.: 9). These are the things that qualify as the ultimate *particulars* or building blocks of the physical structure rather than the things that reveal structure themselves, for example traverse or longitudinal waves, molecules, atoms and the like. Russell says that this does not mean that a particular does not have structure, but only that there is nothing in the known laws of its behaviour and relations that gives us reason to infer a structure (ibid.: 277). He qualifies this again when he makes it clear that having no structure means having no space-time structure, that is parts that are external to one another in space-time (ibid.: 286). Structure in the external parts of events is that to which physical formula refer. Formulas for motion, for example, refer to changes in strings of events.

Events as particulars correspond to the previous conception of simple substances, but since substance has been abandoned as non-empirical, events take their place. For the same reason that substances cannot be basic particulars, events cannot be conceived as points in space nor instants of time, but Russell claims that they occur in less than a second. If they were more than a second long, they would be analysed into a structure of events (ibid.: 287). Thus Russell's events appear to approximate Whitehead's actual occasions, which occur as space-time quanta or epochs that become as whole durations – that is, all at once.

The basis for this ontological conclusion, as we have seen above, is empirical adequacy reinforced by relativity and quantum physics. In the concluding chapter of his *A History of Western Philosophy*, Russell writes:

Common sense thinks of the physical world as composed of 'things' which persist through a certain period of time and move in space. Philosophy and physics developed the notion of 'thing' into that of 'material substance', and thought of material substance as consisting of particles, each very small, and each persisting throughout all time. Einstein substituted events for particles; each event had to each other a relation called 'interval', which could be analysed in various ways into a time-element and space-element . . .

From all this it seems to follow that events, not particles, must be the 'stuff' of physics. What has been thought of as a particle will have to be thought of as a series of events. The series of events that replaces a particle has certain important physical properties, and therefore demands our attention; but it has no more substantiality than any other series of events that we might arbitrarily single out. Thus 'matter' is not part of the ultimate material of the world, but merely a convenient way of collecting events into bundles.

Quantum theory reinforces this conclusion, but its chief philosophical importance is that it regards physical phenomena as possibly discontinuous. It suggests that, in an atom . . . a certain state of affairs persists for a certain time, and then suddenly is replaced by a finitely different state of affairs. (1945: 832–3)

Strings of events connected together give us what we call 'matter'. A light wave differs from a unit of material in that it is spread spherically instead of travelling along a linear path. This also applies to quantum phenomena, especially after Heisenberg, where Russell says the non-substantial character of subatomic particles is revealed even more forcibly ([1927] 1934: 246).[7]

While Russell affirms that we directly perceive events, he maintains that material objects are inferred from sense data; they are logical constructions. Statements about material objects like the moon and the Milky Way galaxy are therefore translated into statements about sense data, much like the manner in which statements about points, lines and planes are translated into statements about classes. To some extent this sounds like the position for which he expresses gratitude to Whitehead, that is the position of making 'the world of matter less remote from the world of our experience', but Whitehead was only concerned to demonstrate how instants of time and geometrical elements could be viewed as logical constructions, not material objects which he views as properties of events. Russell defends this position on the basis of intersubjective agreement among observers as to what they experience, especially when our belief in material objects is based on constant correlations we receive from the different sense receptors, but he is also well aware of problems with this view (ibid.: 197, 204). For one, astronomers see themselves as making statements about heavenly bodies, not sense data. Another considerable difference between Russell and Whitehead's views concerns the manner in which Russell conceives of events on the Newtonian model of the atom. That is, he understands the basic events to be simple, independent and externally related to other events in the space-time manifold. Russell would surely be acquainted with Whitehead's criticisms from the 1920s' books, but he appears to accept the views he is advancing without directly confronting the difficulties that Whitehead was trying to avoid from the outset.

Russell's fully-developed metaphysics is a form of neutral monism in which different arrangements of events or different constructions from them constitute 'mind' and 'matter'. This idea was previously expressed in William James' radical empiricism and it undergoes significant modification in Whitehead's metaphysics. For Russell, the ultimate stuff of the universe consists of neutral events, which are the common ancestors of mind and matter, rather than a dualism of distinct substances, but there is a dualism of causal laws that function very differently in psychology and physics (1921).

QUINE'S EVENTS AND PHYSICAL OBJECTS

Of the revisionary school of thought, Quine has been one of the most recent proponents of an event ontology. As discussed in Chapter 1, he argues forcefully that it is our best theories of cutting-edge science rather than armchair speculation or linguistic analysis that determine our general ontology. The gods of Homer, astrology or any other rival conceptual scheme will not compete as 'best theories' in this sense.

Quine's own attempt to provide a workable ontology that best accommodates science is achieved by accepting a bare minimum of physical objects accompanied by an abstract hierarchy of classes that are based on the physical objects. From the epistemological point of view, physical objects have maintained a dominant status in the organisation of experience. They have survived as our primary objects of reference and they have long functioned as the basis upon which we postulate hypothetical particles. This organisation has been greatly aided by mathematical structures when, appropriately generalised, can be understood in terms of classes.

The motivating force behind Quine's ontological commitment is a rigorous application of Occam's Razor: we postulate entities only to the extent to which we are forced to in the course of explanation. The theoretical economy achieved by doing more with less is a foremost concern in his quest for simplicity and elegance. The key to this strategy is strict regimentation of language that lays bare the ontic commitments of a theory and a programme of ontological reduction wherein certain objects are shown to be derived from the more basic class.

In *Word and Object*, Quine advances this project of regimentation of ordinary language in order to create an extensional language for science. Given the anomalies of ordinary language such as vagueness, ambiguity and various failures of reference, he seeks a regimentation of scientific and ontological language by the purely extensional device of mathematical logic. In accordance with the Einstein-Minkowski concept of space-time, for example, he translates tensed sentences into a canonical notation that construes a specific time as a 'slice of the four-dimensional material world, exhaustive spatially and perpendicular to the time axis' (1960: 172). Physical objects in this world, he says, 'are not to be distinguished from events or, in the concrete sense of the term, processes' (ibid.: 171; also see 1985a: 167). 'A body is thus visualized eternally as a four-dimensional whole, extending up and down, north and south, east and west, hence and ago. A shrinking body is seen as tapered toward the hence; a growing body is tapered toward the ago' (1970: 30). These are the 'worms' of space-time or the 'world-lines' of Minkowski's theory.

So, for Quine, an event is *any* portion of space-time. What he calls a 'physical object' is roughly synonymous with what Whitehead called an 'event', namely 'the material content of any portion of space-time, however irregular and discontinuous and heterogeneous' (1981b: 10; also see 1960: 171). His generalised notion of a physical object or event is intentionally broad so as to accommodate whatever objects are posited by science, including particles, waves, electromagnetic fields and organisms. Quine's broad construal also allows us to think of mass substances, such as the entire world's sugar, sand, dirt and water, as discontinuous physical objects (1981b: 10). Although discontinuous objects of this kind would not typically pass the test for bodies, they can be classified as physical objects in Quine's sense of the term. All of the above comprise the values of the variables of quantification.[8] That is, we commit ourselves to the existence of entities by the use of the bound variable in quantification logic. To be, he says, is to be the value of the bound variable (1953: 12). This criterion does not tell us what there is, but rather what we are committed to saying there is given what our science requires.

Because ordinary bodies fail to provide clear criteria for identify and individuation, Quine says that his liberal notion of a physical object, as being any portion of space-time, spares us this pointless task (1976: 497). Ordinary physical objects are our primary objects of reference, but since the criteria of individuation and identity of such objects over space and time are vague, our concepts of such objects are correspondingly vague. What seem to be self-identical objects soon dissolve into their temporal parts, of which we are equally vague about individuation and identity. Is the desk at a second an entity? What about the desk at a microsecond? Even when we consider a physical body 'frozen' in time, the problem persists. Quine asks: 'Who can aspire to a precise intermolecular demarcation of a desk? Countless minutely divergent aggregates of molecules have equal claims to being my desk' (1985a: 167). And at the level of subatomic particles, our 'robust sense of the reality of physical objects' is disrupted again since identity-conditions fail to provide well-behaved entities. Quine notes that the physicists' age-old attachment to matter has relaxed: 'Matter is quitting the field, and field theory is the order of the day' (1976: 499).[9] It is in relation to field theory that the notion of an event is especially promising as an ontological foundation for physics; the various states are ascribed to regions of the electromagnetic field and bodies seem to fade altogether from the picture. Once again, whether we call this an ontology of 'physical objects', 'events' or 'portions of space-time', it is all the same for Quine.

Quine's very definition of a physical object or event, as 'the material content of any portion of space-time, however irregular and discontinuous and heterogeneous', recognises the fact that there are no precise lines to draw (1981b: 10; 1960: 171). Space-time is 'gerrymandered' in almost any manner desired to suit our purposes. Ordinary language creates a 'vaguely varied and very untidy ontology' that Quine's regimentation seeks to tame (1981b: 9). 'There is room for choice', he says, 'and one chooses with a view to simplicity in one's overall system of the world' (ibid.: 10). His generalised ontology allows us enormous theoretical latitude because it does not restrict us to any specific way in which events or physical objects must be conceived. It is, after all, the business of science, not ontology, to tell us specifically what there is.

As to whether there are functionally relevant groupings in nature in the Whiteheadian sense, Quine's theory is less clear. His discussion of natural kinds seems to point in this direction, given his view that there are both intuitive and taxonomic kinds that are useful for language learning and more general theoretical pursuits, but he refuses to admit properties to his ontology in favour of classes (1969: 114–38).

Physical objects or events are identical, Quine argues, if and only if they are spatio-temporally coextensive. In other words, two events *e* and *e'* are identical when they occupy the same place at the same time. But how do we individuate the identical regions of space-time in the general sense? Quine says that formal logic does the job with the notion of an *extension family* – 'a family of vaguely delimited classes, each class being comprised of nested [events]' (1985a: 168). Each event consists in the activities of some region so that a spatially small event is contained within a Chinese box of larger ones extending outward to the vast event of the whole universe. This certainly sounds like the Whitehead of *The Principles of Natural Knowledge*, but there is a question as to how we make sense of the boundaries of the space-time regions without recognising the existence of properties.[10]

While Quine found much with which to agree in Whitehead's four-dimensionalism, he could not accept the idea that properties are legitimate parts of the ontology of science. Whitehead's ontology is dualistic, containing events and properties (which he called 'objects'). Quine's ontology is also dualistic, containing events (which he also calls 'physical objects') and classes. Whereas Whitehead's 'events' are non-repeatable particulars, his 'objects' are the 'recognisible permanences in nature of various grades of subtlety' (1919: 2); they are the recognisable and repeatable entities in nature. Without objects, as we have seen above, science would be impossible since the laws of nature are discoveries of patterns of objects in the passage of events (ibid.: 87).

In Quine's well-swept ontology, classes were only begrudgingly admitted for services rendered. As he made the point:

> Physical objects in this generous sense constitute a fairly lavish universe, but more is wanted – notably numbers. Measurement is useful in cookery and commerce, and in the fullness of time it rises to a nobler purpose: the formulation of quantitative laws. These are the mainstay of scientific theory, and they call upon the full resources of the real numbers. (1981b: 13–14)

Since numbers are ultimately reducible to classes, the latter are the abstract entities that Quine finally recognises. Like Whitehead, he sees that science would be impossible without them.

Although classes and properties might do some of the same ontological work for science, they are not the same entities. For Quine, however, the legitimate needs that apparently called for properties can be dealt with by classes. In his view, the scientific demand for exactitude requires that we reject properties as bona fide members of our ontology. His stringent requirement for admission is summarised by his criterion, 'There is no entity without identity' (1981b: 102). For any entity considered to be a real object there should be some general criterion of identity for all things of the general kind to which that entity belongs. Classes are identical when their members are identical. But, Quine argues, there is no such clear principle for properties. As he says, if he must come to terms with Platonism, the least he can do is keep it extensional (ibid.: 100).[11]

Whereas Quine identifies physical objects with events in the broadest sense, Whitehead did not consider them to be identical: physical objects are conceived as relatively stable patterns *in* events. Events are primary; objects are secondary. Whitehead's reason for this distinction is based on his claim that events are the entities we directly perceive. This difference in the way Whitehead and Quine understand events is rooted in their different approaches to empiricism.

For Whitehead's theory of natural knowledge, all that there is for knowledge is contained within nature itself. On one front, he argues against the pure phenomenalism of Berkeley by attempting to secure the independence of nature from the knowing mind. On the other front, he argues against the bifurcation of nature into two systems and affirms James' radical empiricism. Perception, he says, is an awareness of events or happenings (1919: 68). 'Experience', means more than just sense experience; it is phenomenologically dense and includes the perception of time as an essential component. In this way, he contends, the immediate perception of events is our most important source of evidence for an ontology of events. The place of the concept of the 'specious present' in Whitehead's theory makes his version of empiricism significantly different from Quine's. They both hold that whatever evidence there is for science is sensory evidence, but 'sensory evidence' for Quine is construed within the context of behaviourist psychology. Defending science from within, for Quine, means that we start with empirical psychology in order to demonstrate how we account for the link between observation and theory – 'between the meager input and the torrential output' (1969: 83). Quine thus takes one step further back from Whitehead's stance in his analysis of sensory evidence, because he starts not with perception but with *reception* construed as nerve endings receiving stimulation.

So, in spite of significant differences between Whitehead's, Russell's and Quine's views on the details of the dualistic ontology, they all affirm a basic commitment to events as the underlying ontology of the physical world.

OBJECTIONS CONSIDERED

Whitehead and Russell argue that an event ontology rather than a substance ontology is more in accordance with the requirements of modern physics and since events are experienced in sense perception, adoption of events as our basic ontology would set physics on a firmer empirical foundation. Events and properties are concrete; everything else, such as the elements of geometry and the structure of space-time, is a derivative abstraction. It is clear, however, that an event ontology constitutes a broad theoretical or conceptual framework. As such, it would seem to

be open to the objection that it offers an interpretation *of* our perceptions. The term 'concrete' in this case functions in the same way as
the term 'real'. As Richard Rorty once put it to me: 'It is abstraction
all the way down.' In other words, since we have no privileged access
to reality, there is no way to establish one conceptual scheme as more
concrete or real above any other (Rorty 1979). All experience is theory
laden. So, in this case, the claim that events are the most concrete entities (as opposed to some other ontological unit) is just to privilege the
event concept at the centre of one's system.

Such anti-foundationalism is a challenge that goes to the very core
of the project of fundamental ontology, but Whitehead and Russell
could defend their thesis by arguing that an event ontology is implied
by modern physics.[12] It is, of course, subject to overthrow in a future
scientific revolution and thereby demonstrates no privileged access to
reality. But that does not mean that any one conceptual scheme is just
as good (or bad) as any other; there is no firmer ground upon which
to stand than that provided by current science. As we have seen above,
Whitehead argues that any observation of nature occurs within the
perceiver's specious present, and within the duration of the specious
present, nature is experienced as events and properties. In an attempt
to avoid 'misplaced concreteness', events provide a stronger empirical foundation than is offered by entities for which we have no direct
experience and, as he and Russell argue, we certainly have no direct
experience of substance as the metaphysical substratum or bearer of
properties, points, instants and so on. Theorise we must; the question
is: Which theory is more in accordance with the science of our time?

A second objection, from a more or less hypothetical critic, might
allege that the difference between a substance and an event ontology is
of no consequence. It is a merely a semantic quibble among academic
philosophers. 'Event' and 'substance' are interchangeable terms for
property bearers; either way, physics will continue to carry on with
absolute indifference to ontology. As noted by Kant (in the quotation
beginning Chapter 2), a debate that rages on in philosophy for a lengthy
period of time is more likely to be a genuine problem rather than merely
a semantic one. In this case, the debate about the nature of fundamental
ontology takes us back to Aristotle if not the Pre-Socratic philosophers.

As we shall see in Chapter 6, there is a tradition of resistance to the
project of fundamental ontology in favour of an instrumentalist or
pragmatic approach, but insofar as a unified theory is concerned, this
tradition is seen as an obstacle to further progress. The concerns of
ontology are far removed from what Kuhn called 'normal science', since
scientists who are engaged in the experimental, everyday laboratory

work are deep into the micro-structure of their paradigm. As such, he claims, they are unaware of the metaphysics implicit to the paradigm (1962). However, since the real issue confronting theoretical physics is the coherence of a comprehensive picture of reality, without which we have merely fragmentary theories functioning independently of one another, the debate does make a genuine difference. Thus, this is not a debate about mere words. It is a debate about how we picture reality as either enduring substances that retain their identity or as dynamic processes that underlie the appearances of endurance. And finally, it is a debate about whether unified theories are superior in their explanatory and predictive power than disunified theories.

5. The Theory of Extension

*We need scarcely add that the contemplation in natural science of
a wider domain than the actual leads to a far better understanding
of the actual.*
(Arthur Eddington)

In this chapter, I explore the connections between Whitehead's metaphysical theory and contemporary cosmology by examining his theory
of extension. This involves the most general features of structure in the
universe that are addressed in his later process ontology. I also continue
to explore the issue raised in Chapter 4 concerning the emergence of
gross physical bodies and the entire extended universe from the foundation of events. As Whitehead's theorising becomes increasingly general,
his account of how our universe began provides a basic framework for
a comparison with contemporary multiverse speculation. This then
raises the question about the scientific status of this hypothesis.

WHITEHEAD'S THEORY OF COSMIC EPOCHS

In *Process and Reality* ([1929] 1978), subtitled *An Essay in Cosmology*,
Whitehead advanced a cosmology as part of his general metaphysics of
process. Metaphysics is the philosophical enquiry into the most general
principles of reality. As he says it is 'the science which seeks to discover
the general ideas which are indispensably relevant to the analysis of
everything that happens' ([1926] 1996: 84) whereas cosmology is 'the
effort to frame a scheme of the general character of the present stage of
the universe' ([1929] 1958: 76). Metaphysics therefore seeks principles
that are necessary for any possible world or cosmic epoch while cosmology discovers by observation what happens to be the case about our
actual world or this cosmic epoch.[1]

As part of his metaphysics, Whitehead formulated a mereological theory that he called 'the theory of society'. This theory of whole-part relations accounts for the order of nature in the extensive continuum. As we saw in Chapter 4, he used the general term 'nexus' to designate a special togetherness of the basic entities of his system. Some nexūs (plural of nexus) are purely temporal or spatial, for example, consciousness and interstellar space. A 'society' is a macroscopic object. It is a nexus that has what Whitehead calls 'social order'. Social order is a common element of form among the entities that belong to any specific society and the imposition of reproduction among the members of that society so that one generation of entities after another reproduce the same pattern. The extended universe is a system of societies embedded in societies embedded in societies. For example, the society of electrons is embedded in the society of atoms, which is embedded in society of molecules, and so on.

This very broad notion of 'society' involves the idea of a character that endures over time given the manner in which the constituent members inherit and modify the defining characteristic. This new metaphysical meaning extends the usual meaning so that a philosophy of process (or an event ontology) now accounts for *things*. A society, for Whitehead, is defined by the massive average objectification of the dominant characteristics or, in his terminology, the eternal objects in the actual occasions forming the society. A structured society is one that includes subordinate societies and subordinate nexūs with a definite pattern of structural interrelations. A molecule, a cell, a planet, a solar system and a galaxy are all examples of structured societies. Most societies with which we come into contact are 'democracies' in the sense that their subordinate societies function together without a central unified mentality. Cell colonies, plants, eco-systems and galaxies are all democracies in this sense. Each society is an organism that is harboured within the environment of another larger society, which serves as an organism for another, and so on. The special sciences – such as physics, chemistry, biology, geology, astronomy – study some layer of society or organisms and their environment – subatomic particles, atoms, molecules, cells, plants, animals, planets and galaxies. Cosmology, the study of the large-scale structure and evolution of the universe, is an investigation of the most general features of organism at the very limits of observation. This is why Whitehead calls his view 'the philosophy of organism' in distinguishing his position from the mechanism of Galileo, Descartes and Newton.

While societies are the things in nature that endure (roughly corresponding to Aristotle's substances), they are not the things that are truly real in Whitehead's ontology. They are rather aggregates of events, the actual occasions in the later theory. Accordingly, subatomic particles, such as electrons and protons, quarks and leptons, or superstrings, would not qualify as the basic entities; they are instead the smallest societies of actual occasions. Change is a character of an event – a nexus of occasions – or a society. Actual occasions, by contrast, become and perish but do not change, again since change, according to Whitehead, is understood as what occurs in a nexus. It is imperative here to note that actual occasions become by prehending other occasions in their immediate causal past. The contemporary occasions forming a society are in unison of becoming and as such are casually independent of one another ([1929] 1978: 123).

Whitehead defines the extensive continuum of the physical universe as: 'one relational complex in which all potential objectifications find their niche . . . it is a complex of entities united by the various allied relationships of whole and part, and of overlapping . . . This extensive continuum expresses the solidarity of all possible standpoints through-out the whole process of the world' (ibid.: 66) from bottom to top, that is, from the most basic entities, actual occasions, which atomise the continuum, to the most general conceivable sort of social order. At the far end, the three largest societies Whitehead postulates within the extensive continuum are: (1) the society of cosmic epochs, (2) the geometrical society, and (3) the society of pure extension. Social order beyond the physical order of any cosmic epoch involves the more general geometrical, mathematical and mereological characteristics, to which any cosmic epoch must conform. As he makes the point:

> In these general properties of extensive connection, we discern the defining characteristic of a vast nexus extending far beyond our immediate cosmic epoch. It contains in itself other epochs, with more particular characteristics incompatible with each other. Then from the standpoint of our present epoch, the fundamental society insofar as it transcends our own epoch seems a vast confusion mitigated by the few, faint elements of order contained in its own defining characteristic of 'extensive connection.' We cannot discriminate its other epochs of vigorous order, and we merely conceive it as harbouring the faint flush of the dawn of order in our own epoch. This ultimate, vast society constitutes the whole environment within which our epoch is set, so far as systematic characteristics are discernible by us in our present stage of development. (ibid.: 97)

For Whitehead, cosmology is the study of the order in our cosmic epoch including the discovery of its general laws. A cosmic epoch is the largest society of events that are governed by a certain set of laws of nature. He says: '. . . the phrase "cosmic epoch" is used to mean that widest society of actual entities whose immediate relevance to ourselves is traceable' (ibid.: 91). More specifically, a cosmic epoch is a vast structured society which includes the vast nexus of interstellar space and the constituent structured galactic societies that exist within a larger geometrical society that itself permits the possibility of diverse dimensionalities of space. Whitehead identifies *our* cosmic epoch with the four-dimensional 'electromagnetic society', of which he credits Maxwell with the discovery of its general character. He says further: 'This epoch is characterized by electronic and protonic actual entities, and by yet more ultimate actual entities which can be dimly discerned in the quanta of energy' (ibid.).

Within the context of post-Hubble cosmology, what Whitehead calls '*our* cosmic epoch' is the electromagnetic society that began at a space-time singularity, known as the Big Bang, roughly 14 billion years ago and that has been expanding and cooling ever since. By the best measurements of astronomers at present, we are able to make observations of the most distant objects now in our cosmic epoch at about 4×10^{26} m away (Tegmark 2007: 99). The sphere of this radius is our *horizon volume* demarcating observable objects from which light has travelled during the 14 billion years since the Big Bang and those even more distant objects that are, in principle, unobservable at the present time.

From the foregoing, it is clear that our universe should be conceived as the entire set of cosmic epochs or cosmoi, one giving birth to another as the order in a predecessor degenerates and gives rise to a new order in a successor. Cosmic epochs, like all societies, arise from disorder (Whitehead [1929] 1978: 91). Our cosmic epoch emerged from the disintegration of its predecessor epoch, and another epoch will emerge from the disintegration of our epoch, perhaps at the moment cosmologists call 'the omega point' at the conclusion of 'the big crunch' or at the rebound initiating a new cycle of expansion. A new cosmic epoch emerges, like the phoenix from the ashes, from the collapse of its predecessor. While Whitehead did not give much attention to the idea of contemporaries, his view of social order beyond cosmic epochs clearly implies that, within the geometrical society and the society of pure extension, there will be a plurality of cosmic epochs, each with a different sort of order and physical laws, existing in causal independence from one another. Our electromagnetic cosmic epoch that is finite and bounded within the wider geometrical society 'constitutes a fragment' (ibid.: 92); others will be characterised by some other general type of order. In other words, Whitehead's cosmoi are spread out in both time and space.[2]

One of the implications of Whitehead's view on the plurality of cosmic epochs is that the laws of nature evolve with the attainment of the ideal for the society in question. In contrast to philosophers, such as Descartes and Kneale, who viewed scientific laws as necessary and universal, that is, omnitemporally and omnispatially unrestricted in scope,[3] Whitehead views scientific laws as contingent, evolving concurrently with the creative advance of nature and restricted in scope to the cosmic epochs in which they apply. He says: 'a system of "laws" determining reproduction in some portion of the universe gradually rises into dominance; it has its stage of endurance, and passes out of existence with the decay of the society from which it emanates' (ibid.: 91). More explicitly, in *Adventures of Ideas*, he says:

> . . . since the laws of nature depend on the individual character of the things constituting nature, as the things change, then correspondingly the laws will change. Thus the modern evolutionary view of the physical universe should conceive of the laws of nature as evolving concurrently with the things constituting the environment. Thus the conception of the Universe as evolving subject to fixed, eternal laws regulating all behaviour should be abandoned. (1933: 143)

The laws of nature are not logically necessary since we can imagine a place where they do not hold without contradiction; nor are they universally and physically necessary since they change with the becoming and passing of the societies in question. Physical laws are grounded in the periodicity of nature that is exhibited within each particular cosmic epoch. Other laws, such as those of geometry and mathematics, hold not only for our cosmic epoch but for all others contained in the geometrical society. Mathematical truths are therefore true in all cosmic epochs. In the most general society of pure extension, extremely general mereological laws apply, such as the relation of whole and part and extensive connection; these will likewise hold in any possible cosmic epoch, but even here Whitehead is hesitant to claim this is a necessary conclusion ([1929] 1978: 35–6). As he emphasises, what we know beyond our cosmic epoch is merely 'a vast confusion mitigated by the few, faint elements of order' (ibid.: 97).

CONTEMPORARY COSMOLOGY AND THE MULTIVERSE HYPOTHESIS

When Whitehead wrote *Process and Reality* in the late 1920s he knew nothing of the great advances made in Big Bang theory, expansion, inflation and the unification of physics that had occurred in post-Hubble cosmology,[4] yet in a rather uncanny manner his theory of cosmic epochs anticipates what has become the most challenging development in contemporary cosmological theory, namely the multiverse hypothesis. His view appears to have support in contemporary cosmology even though his work has not been generally recognised among the more recent proponents.[5]

Contemporary cosmologists, such as Martin Rees, Lee Smolin, Stephen Hawking, Max Tegmark and Steven Weinberg, have speculated that our universe created at the Big Bang is merely one episode, one universe in a multiverse.[6] As Rees puts it in *Before the Beginning*:

> What is conventionally called 'the universe' could be just one member of an ensemble. Countless others may exist in which the laws are different. The universe in which we've emerged belongs to the unusual subset that permits complexity and consciousness to develop . . .
>
> Each universe starts with its own big bang, acquires a distinctive imprint (and its individual physical laws) as it cools, and traces out its own cosmic cycle. The big bang that triggered our entire universe is, in this grander perspective, an infinitesimal part of an elaborate structure that extends far beyond the range of any telescopes. (1997: 3)

The acceptance of such an idea has astounding revolutionary implications for, as Rees argues, the new idea is 'potentially, as drastic an enlargement of our cosmic perspective as the shift from pre-Copernican ideas to the realization that the Earth is orbiting a typical star on the edge of the Milky Way, itself just one galaxy among countless others' (ibid.). If indeed he is right, the multiverse revolution is just as profound a shift in our paradigm as the Copernican revolution was in the seventeenth century. For those who embrace the multiverse hypothesis, it is seen as a plausible step in the progressive enlargement of our understanding, beginning with the geocentric view to the heliocentric view, the galactocentric view to the cosmocentric view, and now to the multiverse.

While cosmologists use different terminology for the multiverse, such as 'megaverse', 'holocosm', and 'parallel worlds', there is general agreement about the need for this new theory that derives from three different but related perspectives: (1) It is necessary to understand the origin of our universe. This is largely due to the attempt to understand emergence from models of expansion and re-collapse. (2) It is also seen as a necessary development of the attempt to find the ultimate unified theory, a Theory of Everything or TOE. One candidate, M-theory, requires extra dimensions beyond the four familiar ones of space and time, of which our universe is merely a 'brane' in a higher-dimensional 'bulk'. Unification is then sought within these higher-dimensional theories. And finally, (3) it gives legitimacy to the anthropic principle, because if there is a multiplicity of universes, it is a simple matter of natural selection that a fraction of these will produce the fine tunings necessary for the emergence of life and consciousness such as we find in our universe.

The two most important 'golden moments' in the development of modern cosmology were Edwin Hubble's discovery of the red shift of distant galaxies and Arno Penzias's and Robert Wilson's accidental discovery of cosmic microwave radiation showering the earth with equal strength from all directions (Peebles 1971). The first led Hubble to formulate his law of kinematics which states that galaxies are receding from us with a speed proportional to their distance. The second, the so-called 'afterglow of creation', is interpreted as leftover radiation from the early hot universe. Both of these discoveries provided observational support for the Big Bang theory and led to the development of the oscillating model, according to which our universe has been expanding and cooling, and will at some point begin to contract to a 'big crunch' where astronomers will begin to observe worrying blue shifts from distant galaxies and the temperature of cosmic microwave radiation will start to rise (Weinberg 1988: 151–2). One simple motivation for proposing the multiverse hypothesis is that it provides a broader theoretical model for understanding how this emergence, expansion and re-collapse is possible. The multiverse hypothesis suggests, contrary to *creatio ex nihilo*, that something had to emerge from something. The Big Bang was the result of the collapse of a predecessor universe.

There are several versions of the multiverse hypothesis, including bubble theory, ensemble theory, string landscapes and Everett's many-worlds interpretation of quantum mechanics. Proponents vary in their interpretations such that some cosmologists view the different universes as spread out in *time*, that is, expansion and re-collapse of a single universe in cycles; some cosmologists view them as spread out in *space*; and others view the universes as spread out in both *time* and *space*. The simplest model of a multiverse, oscillationism, is one that proposes to view the universes as spread out in a single-line succession. This could be conceived as one universe continually undergoing different cycles or epochs of expansion and re-collapse, or a new universe at each fresh Big Bang (Tolman 1934; Weinberg 1988: 153–4). In the bubble scenario, our universe underwent rapid inflation in the early phases of expansion (Guth 1981, 1997), but our universe is merely one bubble among numerous other bubbles that are spread out in space. In the eternal inflation model, which is a combination of the previous two, each universe is continually self-reproducing. According to one version of this theory, cosmologists speculate that when an implosion around a black hole triggers the expansion of a new spatial domain, new 'embryo' universes form within existing ones. From this disjoint space, if those universes are like our own, stars, galaxies and black holes would form and those black holes would in turn spawn another generation of universes and so on (Rees 2001: 158; Hawking 1993: 121; Smolin 1997: 100). We have no information about these universes since we only know our own, but they bear the imprint of their parents, or leave behind an 'umbilical cord' for a baby universe. Again as Whitehead put it, our cosmic epoch is the society that is immediately traceable to ourselves and beyond which we have only faint hints of order.

With regard to unification, one of the most impressive developments has been the string theory scenario which, if successful, promises to be the realisation of Einstein's dream of a unified field theory and beyond by tackling the more difficult unification of quantum mechanics and general relativity (Weinberg 1994: 212). String theory proposes that all matter and all forces of nature are to be understood as a manifestation of particular patterns of string vibrations within multidimensional branes. M-theory, the master theory of all formulations of string theory, or *super*string theory, proposes eleven dimensions (ten of space and one of time). M-theory proponents maintain that our understanding of the electromagnetic space-time continuum (of four dimensions) is merely an evolutionary accident of our sensory organs that has limited our ability to see reality within the boundaries of a narrow band of electromagnetic radiation to which our visual perception is sensitive. In the brane-world scenario postulated in string theory, although 'we could be floating within a grand, expansive, higher-dimensional space, the electromagnetic force – eternally trapped within our dimensions – would be unable to reveal this' (Greene 2004: 393–4). Explaining the force of gravity, however, requires the extra dimensions of M-theory. While at present a work in progress, the expectation is that there will be the development of a theory that will explain the Big Bang by positing a realm in which the cyclic expansion-contraction-rebound occurs. This very roughly approximates Whitehead's notion of the geometrical society harbouring the existence of all cosmic epochs, one that contains all possible geometrical configurations and allows multiple dimensions required by M-theory. According to one version of the theory, our universe is a three-brane set within a string landscape of many other three-branes, all of which are connected and drive the cosmological evolution within the branes by colliding and thereby causing a rebound (Greene 2003: 407). The expansion-contraction-rebound cycle is therefore a result of a much larger cycle of attraction and collision of branes that occurs beyond our universe or cosmic epoch.

The fine tunings of the physical constants (in place and time, in nuclear forces, gravity and chemical elements) that were necessary for the initial emergence of life and consciousness have also contributed to the cautious acceptance of the multiverse concept. The anthropic argument states that the physical constants have certain specific values, when they could in principle have had any values at all. Hence, unless there is a god, there must be an infinite number of other 'places' where the constants take on all possible combinations of values. We find ourselves in one that has become self-aware, because conditions here just happen to be such that consciousness can arise. But if any number of the fundamental constants or initial conditions were slightly different, no complexity would have emerged that would have permitted life and evolution to have taken place. Gribbin and Rees eloquently express this idea about our own place in the scheme of things, as they write:

> . . . we do not inhabit a typical place in the Universe. Most of the Universe is empty space, filled with a weak background sea of electromagnetic radiation, with a temperature of only 3 degrees above absolute zero of temperature, which lies at 273 degrees C . . . Clearly, our home represents a special place in the Universe (although not necessarily a *unique* place). (1989: 6)

One factor that demonstrates the uniqueness of earth is delicate balance of place, the CHZ (continuously habitable zone). If earth's orbit had been only five per cent closer to the sun, the primordial water vapour that was outgassed from volcanoes in the early history of the planet would not have condensed to form the oceans, but rather would have remained in a gaseous state similar to Venus. On the other hand, if the orbit of earth had been even one per cent greater, then the lowered radiation from the youthful sun (coupled with the reduced greenhouse effect) would have left earth covered with massive glaciers in a manner analogous to the deep freeze of Mars (Casti 1990: 351).

Beyond the conditions suitable for life on planets, Rees, in *Just Six Numbers*, has refined the list of finely-tuned cosmological constants to identify which universes provide for the possibility of a *biophilic* universe; any one of these cosmic numbers 'un-tuned' results in a stillborn or sterile universe (2000: 2–3). According to Rees, the requisite cosmological constants needed for the biophilic universe are:

1. $N = 10^{36}$ The measure of the strength of the electrical forces that hold atoms together divided by the force of gravity between them: if N were less, the universe would be too young and too small for life.
2. $\varepsilon = 0.007$ Nuclear binding energy as a fraction of rest mass energy: ε defines how firmly atomic nuclei bind together; if more or less, the complex chemistry for matter and life could not exist.

3. $\Omega = 0.3$ The amount of matter in the universe in units of critical density: if Ω were greater, the universe would have already collapsed; if less, no galaxies would have formed.

4. $\lambda = 0.7$ The cosmological constant in units of critical density: λ controls the expansion and fate of the universe; if it were larger, cosmic evolution would have made it impossible for stars and galaxies to form.

5. $Q = 10^{-5}$ The amplitude of density fluctuations for cosmic structures: if Q were smaller, the universe would be featureless since matter would be blown away from a galaxy instead of being recycled into stars forming planetary systems; if Q were larger, the universe would be dominated by black holes and far too violent for life to exist.

6. $D = 3$ The number of spatial dimensions in our world; if D were 2 or 4, life could not exist. In a three-dimensional world, forces like gravity and electricity obey an inverse square law, which provides for stable orbits of planets around a star and electrons around a positively-charged nucleus.

While the strong anthropic principle asserts that such a universe could only be the product of God or a god, the weak version only requires the more modest postulate of there being a plurality of universes with a variety of properties, of which at least one is hospitable to our existence, that is, the multiverse (Smolin 1997: 203).

AFFINITIES AND CONTRASTS EXPLAINED

The notions of a Big Bang and a dynamic, expanding universe are both consistent with Whitehead's notion of what occurs within a cosmic epoch. Of the diversity of multiverse hypotheses, the oscillationist model – wherein a series of Big Bang to big crunch, expansion to contraction epochs are continually occurring – is the closest fit to Whitehead's model of how cosmic epochs become and perish. The main difference between the oscillationist model and Whitehead's model is that Whitehead thinks that the cosmic epochs are spread out in space as well, each one born from chaos or the disintegration of a predecessor epoch and existing in causal independence of the others.[7] As noted above, there is also a rough correspondence between the larger framework of M-theory and Whitehead's notion of a geometrical society that harbours the existence of cosmic epochs. All of the detailed findings of twentieth-century cosmology are, of course, absent in Whitehead's theory, but disagreements in the variations of multiverse theory aside, the basic insights can be adapted to the general framework that he constructed. The major affinities include the following:

- What we call our universe (or cosmic epoch) is simply one finite element in an infinite ensemble.
- The new universe emerges from the disintegration (or re-collapse) of an older one.
- The laws of nature change from epoch to epoch (or from universe to universe).
- Cosmic epochs (or universes) other than our own are unobservable, but their existence is inferred from what is understood to be a necessary theoretical conjecture.
- In both Whitehead's metaphysics and contemporary cosmology, the multiverse hypothesis was in part a result of a quest for an ultimate unified theory.

Whitehead continued to use the term 'universe' to refer to the totality of what is, namely the whole scheme of the extensive continuum, whereas contemporary multiverse theorists refer to what he calls a 'cosmic epoch' as one universe among a multitude of universes. What was previously believed to be the totality has been surpassed by a new theory; both Whitehead and contemporary cosmologists have contributed to this enlargement of thought about the structure of reality.

As for the contrasts between Whitehead and contemporary multiverse theorists, there is one major difference that pertains to the laws of nature. Whitehead thinks that the fundamental laws of nature change from epoch to epoch. In most contemporary views, by contrast, there must be one basic, unified theory of physical laws that governs all universes in the multiverse, but that gives rise to different local laws due to different outcomes of symmetry breaking. Rees, for example, regards the exact layout of planets and asteroids in our solar system or even the structure of galaxies in our universe as accidents of history and therefore arbitrary, so *'the underlying laws governing the entire multiverse may allow variety among the universes'* (2001: 173). The local by-laws are variations of the more stable laws in the grander perspective. The search for a Theory of Everything presumes a fundamental stability in the ultimate laws even if there is local variation.

This difference regarding the laws of nature can be explained as a difference in ultimate metaphysics. Whitehead's metaphysics is one in which process is the fundamental principle of reality; physical laws change with the advance of novelty. Contemporary cosmologists and physicists seldom, if ever, explicitly express their views in metaphysical terms. Nonetheless, insofar as there is some basic ontology underlying multiverse views, they appear to embrace the classical field theory as a rough model for the Theory of Everything and they appear to interpret the field as a type of substance which remains the same throughout change. This ultimate substance functions as a kind of matrix for the becoming of the individual universes. The dispositional properties determine the varying intensities of the field but the field itself provides for a stability in which cosmologists seek the fundamental laws. M-theory, for example, conjectures a type of receptacle constituted by a higher-dimensional Superspace in which the different universes emerge. Even in the case of the oscillating model of a single-line succession of universes, there is still the notion of persistent laws throughout the collapse of predecessor universes. Some physicists think that the different forces – gravitation, electromagnetism, weak and strong nuclear interactions – are united at extremely high temperatures (10^{32} ° K) and become differentiated through spontaneous symmetry breaking as the temperature drops (see Weinberg 1988: 145–6; Rees 1997: 152–3). The outcomes of the symmetry breaking might be different on different occasions, but statistical predictions from the basic unified theory would describe the different local by-laws in each rebound (see Hawking 1993: 63). The Theory of Everything seeks the ultimate laws in the ensemble or infinite sequence that remain unchanging. In this way there is a lawful change of the local by-laws in each universe.

On this point it might initially seem that Whitehead's view will be unacceptable to most modern cosmologists since it appears to be in violation of the basic premise behind the search for a Theory of Everything. We should, however, keep in mind that he explicitly made provisions in his theory for the stability of the more general laws. In the hierarchy of social order whereby the societies of cosmic epochs are harboured within the geometrical society and, beyond that, the society of pure extension, the stability of more general laws is to be found in these larger societies. Whitehead's point is that the nature of the laws must be based on the nature of the things to which they refer; if the things change, then so do the laws, but as we ascend the extensive continuum the change in the wider societies is so miniscule that the laws at that level will appear to be permanent. The problem, however, is that these laws are mathematical, geometrical and mereological laws rather than physical laws of nature. Rees, who argues for a complementarity between chance and necessity in the laws governing the production of the individual universes, nonetheless concedes that there is no consensus among physicists on the issue: 'there could be a unique physics;

there could, alternatively, be googles of alternative laws' (2007: 65). In this respect, Whitehead's view is not ruled out.

CRITICAL EVALUATION

Cosmology has now arrived at a critical juncture. The multiverse hypothesis seriously entertained by the mainstream of physics challenges our traditional conception of science in which the necessity of observational or experimental corroboration has been emphasised. Detractors therefore argue that the multiverse hypothesis is: (1) wildly speculative, posits unobservable entities and is untestable, and (2) a violation of Occam's Razor in that it multiplies entities (or universes) beyond necessity and results in an aesthetically ugly theory. If universes or cosmic epochs are disconnected from our own and therefore are unobservable in principle, then there can be no empirical means of corroboration. Some physicists have charged that it is not science at all, but rather metaphysics (Davies 2007: 491). Proponents counter that there is a price to pay for forswearing cosmological speculation. Our hubris and lack of imagination in the past have been obstacles to scientific progress. In this connection the dismissal of the multiverse hypothesis on the basis of our failure to make the necessary observations *from our vantage point* shares far too many similarities with previous episodes in the history of physics where the vastness of the physical world has been underestimated (Tegmark 2007: 100).

In the middle ages it was inconceivable to most people that there could be other solar systems in the universe. Giordano Bruno, however, offered the bold conjecture that there are other solar systems (as well as other universes) and became the martyr of science for his heretical opinions. Commenting on Bruno's fate, Whitehead writes:

> the cause for which he suffered was not that of science, but that of free imaginative speculation. His death in the year 1600 ushered in the first century of modern science in the strict sense of the term. In his execution there was an unconscious symbolism: for the subsequent tone of scientific thought has contained distrust of his type of general speculation. (1925: 1–2)

Just as the exoplanet hypothesis was at one time a philosophical speculation which through technological progress became a scientific speculation, the multiverse hypothesis could very well run the same course. It might very well be inconceivable to us at present what future technology could possibly advance multiverse speculation into the realm of the empirical but so it was as well for the medieval thinkers with respect to exoplanet discovery by radio and space telescopes.[8]

Given Whitehead's very definitions of metaphysics and cosmology it is clear that the highly speculative multiverse hypothesis falls within the province of metaphysics, but since he never embraced a sharp dichotomy between metaphysics and science the scientific status of the theory of cosmoi was not a matter of concern. As we saw in Chapter 1, Whitehead rejected the traditional concept of metaphysics as a purely *a priori* endeavour for the idea that it is the general end of theory that originates in natural science. Science begins in the general description of observed fact, but Whitehead notes that the impulse towards speculation is grounded in the unrest with which scientists are confronted. Hubble, for example, thought that our knowledge fades rapidly with increasing distance; when we have exhausted our empirical resources, the limits of our telescopes, etc., we then pass into the 'dreamy realm of speculation' (1936: 202). Dreamy or not, the speculation is necessary. Lack of satisfaction with simple description or even the *general* description of observed fact is the justification of speculative extension. This urge towards an *explanatory* description is the basis for the continuum between science and metaphysics (Whitehead 1933: 164). In *The Function of Reason*, Whitehead describes the necessity of metaphysics as arising from the fatigue of methodology and the need for refreshing novelty in answering fundamental questions ([1929] 1958: 22–3). Positivism is just one example of methodology, of strict adherence to observation, which the history of science itself demonstrates is untenable.

The evidence for other cosmic epochs is indirect; it is at best an inference from what is observed in our cosmic epoch. Thus the hierarchy of societies beyond our cosmic epoch must be considered metaphysical in the sense that it is an extension of theory beyond the observable, yet it is clear that our epoch must be set in a larger society that serves as its environment and it must have originated from the disintegration of its predecessor epoch. Whitehead thus argues that the inference is justified by the search for a more complete theory. He described his theory as a 'conjecture' ([1929] 1978: 96), but it appears to be a conjecture that is necessary to achieve the goal of complete unification. This, I take it, is what he means by the urge toward *explanatory* description taken to its ultimate conclusion.

The sharp boundary between metaphysics and science on the basis of observability is faced with an undeniable difficulty given the strong physical evidence for an expanding universe. What is observable is limited by two horizons: *the technical horizon* determined by the sophistication of our telescopes at present and *the speed-of-light horizon* determined by the present rate of acceleration (Rees 2007: 61).

With the development of more powerful telescopes in the future more galaxies will be revealed, and as this happens our present technical horizon will be extended. If, however, the expansion of the universe continues to accelerate, light from the most distant galaxies will never reach us and, as a consequence, these most distant galaxies will, in principle, remain unobservable to us; but if the expansion decelerates at some point, resulting in re-collapse, those galaxies will be visible in a very remote future. Our cosmic epoch (or universe) extends beyond our present 10–20 billion-light-year horizon; this is not metaphysics but rather an inference from observed fact. The inference to predecessor and contemporary universes is admittedly greater, but the objection that they are ruled out by the observability criterion loses all force when we realise that the boundary between observable and unobservable is blurred by the dynamics of expansion.

Even Karl Popper, whose criterion of falsifiability would rule out the multiverse hypothesis as science, recognised that metaphysics is an inevitable precursor to science, as a sort of embryo in the development of scientific hypotheses (1992: 199–211). The question remains, however, as to whether the multiverse hypothesis is testable and therefore falsifiable. Is there any conceivable test in which the hypothesis could be refuted? As expected, the physicists at present disagree on the answer to this crucial question, but as theory guides experiment and experiment further refines theory there are hints of what future experiments can be conducted. Just as Einstein's theory of general relativity took more than fifty years before any reliable tests could be conducted that gave results with better than ten per cent accuracy, twenty-first century physicists have outlined generally the sorts of testing that could in principle falsify particular multiverse hypotheses (Rees 2001: 171–2, 2007: 66–74).[9] The multiverse hypothesis is not part of the fundamental physical theory that is accepted at present, but it is not irredeemably untestable; as a speculative idea that is currently considered to be a solution to many problems it could join the ranks of accepted theory if it becomes testable.

With regard to the second major objection to the multiverse hypothesis, the history of physics has demonstrated that simplicity in theories is the result of successful unification, as for example in the case of Newton's law of gravitation and Maxwell's equations of electromagnetism. Multiverse theories seek the same result in unification, but pay a high price in ontology in a manner that is analogous to set theory in mathematics. Whitehead's famous quip on the subject is instructive: 'The guiding motto in the life of every natural philosopher should be, Seek simplicity and distrust it.' He argued that every age prides itself on having discovered 'the ultimate concepts in which all that happens can be formulated', but the problem is that we fall into the trap of 'thinking that the facts are simple because simplicity is the goal of our quest' (1920: 163). Occam's Razor remains a guiding principle for the development of eloquent and beautiful theories, but not to the point of restraining speculation when the sheer complexity and magnitude of things suggest otherwise. Whitehead was always clear that reality is complex beyond our apprehension and our ability to express our apprehension; oversimplification is the ever-present danger in both philosophy and science.

Again given Whitehead's definitions of 'metaphysics' and 'cosmology', process and the creative advance of nature would apply to any cosmic epoch or possible world since these are the ultimate principles of his metaphysics. In his view, the quest for a grand unified theory would apply only to our cosmic epoch – a mere phase of the universe that began with the Big Bang – but a Theory of Everything would apply to all cosmic epochs.[10] It might appear that the ontology of events sketched in our previous chapters would only describe *our* cosmic epoch since it is natural science that determines the character of our general ontology and the empirical evidence supports the idea that events with an electromagnetic character are basic in our cosmic epoch. However, if on a grander metaphysical scale process is the ultimate principle, as it is in Whitehead's system, then the cosmic epochs are the very largest events on a cosmological scale, of which physics describes the general social order. Like the actual occasions, cosmic epochs emerge from their predecessors, are independent of their contemporaries and will perish in order to provide the raw material for their successors. Thus Whitehead must view the event ontology as part of the ultimate metaphysics describing the process of all cosmoi, but the character of those societies would change from epoch to epoch. Contrary to the title of the present book then, it is an *Event Multiverse*, rather than an *Event Universe*, that describes the full scope of the event ontology.

Whitehead's theory of cosmic epochs might very well appear quaint to the physicist steeped in contemporary string theory or inflationary cosmology. There is no doubt that his theory offers little in terms of detailed science or a fruitful direction for testing multiverse theory. The point, however, is not what Whitehead contributes today, but rather how well he pioneered a general framework for multiverse theory roughly seventy years before such theories began to enter the mainstream of physics. Whitehead's genius lies in his power of generalisation, particularly in his formulation of a metaphysics from the early advances in twentieth-century physics. He saw quite early in the game

that the breakdown of the Newtonian paradigm required a new unifying concept to bring together the fragmentary theories of physics,
from the large-scale structures determined by gravity to the small-scale
energetic vibrations occurring at the quantum level. Part of this project
involved his theory of cosmic epochs and a process metaphysics that
explained their emergence and decay. Many physicists have become
champions of Whitehead's philosophy for the general frameworks, the
ontological foundations and unifying concepts he provided rather than
the particular details. Whitehead also weighs in on the contemporary
debate concerning the scientific status of the multiverse conjecture.
Physics without speculation is sterile. Some metaphysical daring is
required to break the spell of custom and conjure fresh perspectives
– ones that will need to be formulated specifically and result in the possibility of testing to be taken seriously.

Plato famously said in his masterpiece of cosmology, *Timaeus*, that
any account of the cosmos is at best a likely story. That is, as mere
mortals 'we ought to accept the tale which is most probable and inquire
no further', for the universe is a process of becoming and so are our
accounts of it (1953: 717). Indeed the very idea of a final theory in
physics is a receding horizon perhaps for the very reason Plato gave.
It is in this connection that Whitehead praised the *Timaeus*: 'what it
lacks in superficial detail, it makes up for by its philosophical depth'
([1929] 1978: 93). Multiverse theory, whether that postulated by
Whitehead or contemporary cosmologists, might very well be a likely
story, but in our quest for enlarging our understanding by theoretical
unification it appears to be a rational development even if disparagingly
metaphysical.

6. The Problem of Time

> *The distinction between past, present, and future is only a*
> *stubbornly persistent illusion.*
> (Albert Einstein)

Leaving aside the multiverse hypothesis, M-theory and superstring theory as possible avenues for some 'ultimate' unification, let us now focus on the difficulties in producing a unified theory for *this* universe or cosmic epoch. If we accept events as our fundamental ontology, we have a basis for systematically interpreting both microscopic entities such as subatomic particles and macroscopic entities even at the largest cosmological scale in terms of events. This has been the upshot of our thesis thus far. Eliminating substances in favour of events, however, does not get us very far towards a unified theory. It is only the beginning, albeit a critically important one.

In this chapter, I examine the problem of time with regard to the unification of general relativity and quantum mechanics and advance in broad outline one direction that this unification might potentially follow, namely a version of C. D. Broad's growing block universe consistent with Whitehead's late metaphysics and relativistic quantum field theory. One of the great challenges of unification, as noted in Chapter 1, is to solve the problem of the very different conceptual roles that time plays in general relativity and quantum mechanics. Moreover, it is not at all clear how we are to reconcile the geometry of space-time within which each event has its own light cone and other obvious physical phenomena such as becoming and temporal succession. Whitehead certainly struggled with this problem when he moved from a purely four-dimensional system to one in which time plays a more fundamental role, that is a temporalisation of space rather than a spatialisation of time, and finally developed his solution by embracing a fundamental asymmetry of time in his metaphysics of process. Other philosophers and physicists have followed Whitehead's lead in formulating tentative solutions in which general relativity is modified in order to account for the creative advance of nature.

Aside from the problem of time, another major stumbling block to unifying general relativity with quantum mechanics is the failure to develop an overall picture of reality. On the one hand, general relativity describing the gravitational force is of deterministic form for all events – past, present and future – ontologically fixed in four-dimensional space-time. On the other hand, quantum mechanics provides statistical laws that describe the outcome of performing measurements on quantum systems. Proponents of the orthodox view contend that everything proceeds deterministically until a measurement is made which then gives us a probabilistic view about future events *but without any view supposed about the nature of the underlying reality*. So, general relativity is realistic in its physical ontology whereas orthodox quantum theory is purely instrumentalist for its lack of any such ontology. This conflict must be resolved in a final, unified theory.

INSTRUMENTALISM AND QUANTUM ONTOLOGIES

The terms 'orthodox theory' and the 'Copenhagen interpretation' are used to describe Bohr's and Heisenberg's description of nature that resulted from their belief that the explanation of the quantum world required a radical departure from classical physics. This radical departure includes: Planck's discrete quanta explanation for the blackbody phenomenon, Bohr's principle of complementarity to explain the wave-particle duality, and Heisenberg's uncertainty principle which describes the restriction on measuring conjoint variables such as position and momentum. Whereas classical physics appeared to be compatible with a causal view of nature, according to which all that happens in nature is, in principle, predictable since every definite cause gives rise to a definite effect, quantum mechanics has found that, at the level of subatomic phenomena, there is an irreducible uncertainty involved in predictions. As Walter Heitler, the author of *Elementary Wave Mechanics*, writes:

> Whilst Newton's equation of motion allows one to calculate the orbit of a particle for any time accurately (in any given circumstances), the wave equation only allows one to calculate ψ, i.e. the probability for finding the particle at a certain position. This replacement of accurate and predictable orbits by uncertain values and probability distributions is the chief step in the transition from classical physics to wave mechanics. (1945: 18)

According to the Copenhagen interpretation, fiercely debated at first but then generally accepted by physicists, the description of nature at the atomic level is derived from mathematical probability functions that describe the possible results of observation rather than classical objects. As Bohr makes the point: 'In our description of nature the purpose is not to disclose the real essence of phenomena, but only to track down as far as possible relations between the multifold aspects of our experience' (1934: 18). And again, 'the formalism does not allow pictorial representation along accustomed lines, but aims directly at establishing relations between observations obtained under well-defined conditions (1958: 71).

In such fashion, Bohr articulated what became the orthodox view of quantum mechanics, and, more generally, a rigorous scientific attitude against any speculation into the micro-structure of nature. This orthodox view is primarily an *epistemic* stance that is closely related to the positivistic rejection of metaphysics and the pragmatic or instrumentalist emphasis on the practical success of our ideas in experience.[1] It contrasts with the *ontic* stance or the view that adopts a robust realism with respect to atoms, fields and subatomic particles postulated by classical and relativistic physics.[2] Bohr's pragmatic interpretation of quantum phenomena has been enormously successful as a method of calculation and prediction, but as a set of rules for calculating correlations among observations it provides no genuine insights about the ontological underpinnings that exist at the quantum level. Proponents of orthodox quantum theory, therefore, accept a fundamental positivistic belief that ontology lies outside the scope of science. As we shall see, however, this epistemic stance of orthodox quantum theory is fraught with serious conceptual problems in spite of its great success.

Schrödinger's wave function Ψ is central to the Copenhagen interpretation. It is understood as the probability that a concrete particle is likely to be found at some location. Subatomic particles have *tendencies* to exist rather than definite locations at definite times. These tendencies are expressed as probabilities that take the form of waves – that is, abstract probability waves. The evolution of the wave function is the smooth development of the wave as it continuously spreads out in different directions and the collapse of the wave function is the manifestation of a physical particle. The Copenhagen theory states that the quantum state of a physical system is described by the wave function Ψ which is defined relative to: (1) some preparation procedure, and (2) some process of measurement. The collapse of the wave function is the point at which a measurement is made in the system. A central part of this interpretation involves the notion that what is being observed and the set-up (including human observers or consciousness) are critical to the appearance of the particle. The observed and the observer are inseparable. In fact, according to this theory, observations produce what is measured. This is yet another radical departure from classical physics where context is normally ignored and the scientist is understood to be a detached observer exploring an independent and objective reality.

The Copenhagen interpretation has been hotly contested for a number of reasons. Nicholas Maxwell, for example, considers it to be a seriously defective theory since it is 'imprecise, ambiguous, ad hoc, non-explanatory, restricted in scope, and resistant to unification' (1988: 3–7; 1994: 351). In short, he argues that it fails as a complete, comprehensive theory in spite of the fact that it forms one of the most important components of the standard model of particle physics. There is perhaps no better place to see the quandary presented by the epistemic stance than Bohr's principle of complementarity as a workable basis upon which to understand the quantum formalism. The problem is that it dodges the wave/particle problem rather than provides a solution for it. Maxwell charges that Bohr, Heisenberg, Born, Dirac and others failed to acknowledge the fundamentally problematic character of the quantum domain and thus failed to develop quantum theory as a fully realistic theory (1988: 1–2; 1994: 352). They also developed the theory in such a way that it is restricted to predicting the outcome of performing measurements on quantum systems. This has the baffling consequence that the occurrence of events in nature is dependent upon our presence and specifically the activities of physicists who perform the measurements (1998: 234). Quantum systems such as electrons, protons, photons and atoms are, for Maxwell, inherently probabilistic physical entities and so he maintains that the solution to the wave/particle dilemma is to specify precise quantum mechanical conditions for probabilistic transitions but, in contrast to the Copenhagen interpretation, without reference to measurement (1994: 352).[3] It is precisely this probabilistic character of the quantum world, together with his plea for realism, that will become an important part of the view that will be proposed in this chapter.

Another considerable problem with the Copenhagen interpretation is the fact that it isolates the quantum system from the wider environment. Henry Stapp argues that there is a fundamental incoherence in this isolation because the Copenhagen interpretation presupposes that the quantum system is embedded within the objects of the classically-described world in which it is situated. The interactions between the quantum system and the measuring system (including the observer) are such that it is impossible to consider the quantum system as separately existing, yet it must exist separately in order to represent the quantum system by a wave function governed by the Schrödinger equation (Stapp 1979: 10). While this might be pragmatically justified, the Copenhagen interpretation faces the inevitable consequence that the major scientific advances have come historically from the successful unification of physical descriptions in wider domains of experience rather than from settling for fragmentary theories. But the conception of an isolated quantum system excludes a unified description of phenomena as for example in molecular biology where the exchange of matter between the system and the surrounding environment is essential to understanding the phenomena under investigation.

This 'classically-described world' in which observations are made includes a Newtonian conception of time, assumed, but not acknowledged, in the theory.[4] That is, there is certainly a presumption that time is ticking away in a manner that is independent of the quantum system, yet this presumption is not recognised nor integrated into the overall theoretical framework of the Copenhagen interpretation. The same goes for space, which is assumed to be the environment that contains the quantum systems that are under study. All of this, of course, clashes with relativity theory wherein the concepts of space and time are intertwined.

As much as proponents of the Copenhagen interpretation resist any ontological assumptions about nature, they do so only through the artificial isolation of the quantum system. As is typical of positivistic schemes, however, the epistemological imperative to resist descriptions of the underlying reality in the course of obtaining data from strict observations runs into the problem of assuming basic concepts in the course of making the observations. Just as logical positivism failed to eliminate metaphysics from scientific enquiry, the Copenhagen interpretation failed to produce a theory that avoids ontology and the broader metaphysical assumptions about space, time and measurement. On this score, Heisenberg insists that in science one is not interested in the universe as a whole, but only in some part that is the object of one's studies (1958: 52). But as Stapp rightly maintains, in order to understand quantum theory one must see that the quantum system, the surrounding environment and the observer-scientist are all 'dynamically linked' (1979: 10–11).

So, the first step towards a unified theory requires a rejection of the instrumentalism of the Copenhagen theory so that a consistent realism across the theoretical framework and an enlargement of the scope of quantum theory might be achieved. If quantum theory is meant to be understood merely as a set of computational rules, then speculation about the way the universe is such that the rules hold in our universe cannot be useless. Optimal use of the rules is impaired if they are disassociated from a coherent conception of that to which they are meant to apply.[5] This leads us to consider quantum ontologies, of which, according to Stapp, the main contenders are: (1) the pilot-wave ontology, (2) the many-worlds ontology, (3) the spontaneous-reduction models, and (4) the actual-events ontology (2011: 55–98). For obvious reasons, we are here interested in the actual-events ontology, a synthesis of Whitehead and Heisenberg, which Stapp considers the simplest alternative to the other 'metaphysically extravagant' theories. Numerous theorists have, in fact, found Whitehead's ontology to be an intelligible model for understanding the quantum world.[6] Stapp says it provides a 'natural theoretical framework in which to imbed quantum theory' and contrasts the pragmatic approach of the Copenhagen theory with Whitehead's ontological interpretation (1979: 1).

In working towards the goal of a comprehensive, unified theory Stapp has therefore developed a Whiteheadian quantum ontology that incorporates some of Heisenberg's main principles. He has argued that Whitehead's ontological approach accords with Einstein's view that physical theories should refer normally to the objective physical situation rather than to our knowledge of that system (ibid.). His main reasons for advancing this synthesis include the following: (1) Whitehead's actual occasions are functionally similar to the collapses of wave functions that play a key role in quantum mechanics. In this model, the wave function is not merely a tool for calculating correlations among observations but rather is treated as an appropriate mental representation of the world itself. (2) Both Whitehead's ontology and orthodox quantum theory propose that these events have a *mental* aspect and that they are causally efficacious in the physical world. In orthodox quantum theory, this causal efficacy of experiential realities arises from von Neumann's 'Process 1', which injects effects of conscious choices crucially into the dynamics. The 'choice on the part of the experimenter' described by Bohr and Heisenberg is the manifestation in the physical world of this choice. In Whitehead's theory, however, it should be clear that the 'choice' is what is occurring in nature itself, not the experimenter. That is, the choice is the activity of actual occasions prehending their predecessors. (3) Heisenberg, when considering what is *really happening*, speaks of these reduction events as follows:

> [T]he transition from the 'possible' to the 'actual' takes place during the act of observation. If we want to describe what happens in an atomic event, we have to realize that the word 'happens' can apply only to the observation,

not to the state of affairs between two observations. It applies to the physical, not the psychical act of observation, and we may say that the transition from the 'possible' to the 'actual' takes place as soon as the interaction of the object with the measuring device, and thereby with the rest of the world, has come into play; it is not connected with the act of registration of the result by the mind of the observer. The discontinuous change in the probability function, however, takes place with the act of registration, because it is the discontinuous change of our knowledge in the instant of registration that has its image in the discontinuous change of the probability function. (1958: 54–5)

While Heisenberg does not appear to have changed his view about the underlying ontology of the quantum phenomena, he has modified his earlier view by adopting a quasi-Aristotelian concept of potentia.[7] The evolution and collapse of the wave function is here characterised as a transition from potential to actual events, or as he put it: the Aristotelian potentia 'introduced something standing in the middle between the idea of an event and the actual event, a strange kind of physical reality just in the middle between possibility and reality' (ibid.: 41).

The weirdness of the quantum world that Heisenberg describes as being a 'strange kind of physical reality' bears a certain affinity to Whitehead's description of epochal becoming via the mechanism of prehension. So, when Stapp says that Whitehead's actual occasions and the collapse of the wave function are 'functionally similar', he means that both of these ideas propose to explain how the discrete emerges from the continuous, namely a particle emerges from a continuous quantum smearing of possibilities. As Whitehead put it, actual occasions 'make real what was antecedently merely potential' ([1929] 1978: 72). According to Stapp, the wave function of the quantum theorist is a mental counterpart of an 'absolute wave function' that represents potentialities. Each actual event is represented by a 'quantum jump' in the absolute wave function. In this conceptual framework, the basic process in nature is a sequence of events, each of which transforms the tendencies created by prior events. He describes this as a stochastic system, one in which given any present state a subsequent state of the system is determined probabilistically.

Stapp's adoption of Heisenberg's concept of potentia together with Whitehead's notion of prehension provides a description of nature at the micro-level. This is something, Stapp claims, none of the founders of quantum theory had done (2011: 89). According to this model then, electrons, protons and neutrons behave in ways that are something like the epochal becoming of actual occasions. They exist as *fields of potentialities*. Just as the occasion only appears when its process of becoming is complete, the subatomic particles become fully actualised when some effective observation is made. In other words, given the non-classical behaviour of the microscopic quantum world, Stapp finds in Whitehead's actual occasions a model for picturing the quantum world. But it should be clear that Whitehead's 'actual occasions' and Stapp's 'actual events' are not identical terms. For Whitehead, we must recall, actual occasions are understood to be atomic events that comprise the plenum, including the vast nexus of 'empty space'. Subatomic particles, in his view, are societies. This means that Stapp's actual events would count as societies in Whitehead's ontology.[8] For both Whitehead and Stapp, however, the central point is that quantum phenomena behave more like events smeared out over space-time than the enduring objects of classical physics. So, at the rock-bottom ontological level, the becoming of actual occasions is basic, and at the first level of physical integrative activity where subatomic quantum events occur there is greater similarity with the actual occasions than with the classically-described macroscopic objects of ordinary perception. Stapp is after all producing a quantum ontology. For him, Whitehead's metaphysics helps us understand how the anthropocentric and pragmatic approach of the Copenhagen interpretation can be replaced with a conception of natural processes that enables us to see the human involvement speci-fied in quantum theory in a non-anthropocentic manner while retaining the essential innovations of the theory. In this way, some of the seem-ingly bizarre features of the Copenhagen interpretation are given a realist, albeit metaphysical, grounding.

In accordance with Whitehead's ontology, Stapp's actual-events ontology modifies the typical conception of time in quantum mechan-ics. Instead of conceiving of time as existing external to the quantum system, he proposes that it should be conceived as an internal part of quantum systems – as coextensive with quantum processes. Quantum phenomena are essentially temporal events as the merely potential becomes actual in successive collapses of the wave function. Instead of thinking of quantum events as happening *in* time, they are the basic units *of* time, each a 'slab' of time-space. Physical time and physical space emerge from the dynamic activities of micro-events, which are themselves units of time-space.

In his *Dreams of a Final Theory*, Steven Weinberg writes: 'If there is anything in our present understanding of nature likely to survive in a final theory, it is quantum mechanics' (1994: 66). No doubt the experimental findings and much of the formalism of quantum mechan-ics will survive, but not necessarily the Copenhagen interpretation, which is notorious for making the quantum world incomprehensible. Stapp has cogently argued that this will not serve physics in the long run. Quantum mechanics is an essential part of the final unification but only if we acquire a coherent understanding of what is actually going on in the collapse of the wave function and how the quantum domain is integrated in the larger environment (2009: 322; 1979: 12). That which remains puzzling and disturbing in the Copenhagen interpretation becomes intelligible when we adopt a new paradigm that illuminates the physical situation.

SPECIAL AND GENERAL RELATIVITY RECONSIDERED

The actual-events ontology involves the idea of a fixed past and an open future, the latter constrained by activities of the present, but this indeterminism appears to be incompatible with the determinism of special and general relativity. Thus, just as quantum theory gets modified in the development of a unified theory, so does relativity theory. As many physicists have argued, we need to give up the idea of starting with an *a priori* given space-time manifold in order to make any progress at all towards a unified theory. The geometricised, four-dimensional view of space-time in which time behaves more like space is the focus of this modification. That is, given the multiplicity of inertial frames of reference within the space-time manifold, each with its unique temporal sequence of events and a different relation of simultaneity, time in the standard approach is frozen and becoming is treated as illusory. But for the unified theory proposed here, becoming is treated as the primary principle of reality in explaining the creative advance of nature.

This move was anticipated by Milič Čapek who devoted much of his work to defending the reality of temporal becoming against what he called the 'fallacy of spatialization' in the Einstein-Minkowski model of space-time.[9] For Čapek, relativity theory need not be interpreted according to an Eleatic tradition of viewing the world as a timeless, four-dimensional entity. But since the theory of gravity in Einstein's general relativity is seen as a firmly established physical theory, most physicists will be resistant to the idea that its rejection is the key to the final unification. If, however, the standard interpretation of Einstein's theory is modified without necessarily involving any direct changes to his formalism, one major stumbling block towards a potential unified theory would be removed. Quantum mechanics and a theory of gravity can be united by a process model of time that preserves the distinctions between past, present and future and views extension as a self-organising system emerging from dynamic temporal processes.

Stapp's synthesis, as I shall call it, attempts this unification via the

Whitehead-Heisenberg interpretation of quantum phenomena supplemented by developments in relativistic quantum field theory. Stapp argues that certain key features of relativity theory must be altered in order to develop an open future theory. This was accomplished by Sin-Itiro Tomonaga and Julian Schwinger in the 1940s and 1950s.

In non-relativistic quantum theory, there is an advancing constant-time surface that demarcates the already determined past from the open future. Quantum theory thus adopts the classical idea of instants in which quantum collapses occur. In von Neumann's interpretation, for example, each quantum reduction event produces a new quantum state $\Psi(t)$ of the universe at the instant denoted by the time t. Each occurs at a particular 'now' but extended throughout all of space. The advancing surface builds upon all the subsequent surfaces constituting the fixed past in a linear sequence in what looks like layers of a cake, one temporal layer at a time. Tomonaga and Schwinger relativised this idea to produce relativistic quantum field theory. The constant-time surfaces of the non-relativistic theory are generalised to space-like surfaces that can be formed by continuous transformations of a constant-time surface that maintain the condition that every pair of different points on the surface are separated by a space-like displacement. The unison of becoming is tied to a single advancing space-like surface in which potentialities become actualities. $\Psi(t)$ is then replaced by $\Psi(\sigma)$ where σ denotes a continuous three-dimensional surface in the four-dimensional space-time continuum. Every point on that surface is space-like separated from every other point (Stapp 2011: 92–4).[10] But the crucial idea, as Stapp put it, is that:

> The advancing succession of flat 'instants' of the non-relativistic theory, upon which the collapses occur, are replaced by an advancing sequence of space-like surfaces, where 'advancing' means that every point on one space-like surface either coincides with a point on that preceding surface but inside the open forward light cone of some of the points on that preceding surface: the succession of space-like surfaces upon which the collapses occur advance *locally* into the future. (2009: 337)

Stapp notes the agreement of these ideas with Whitehead's philosophy of organism wherein the fixed and settled facts grow via sequences of actual occasions. In effect, the development of nature resembles cell growth except that instead of developing within a pre-given space-time, it is time-space itself that is developing. Objectified past actualities become data for prehensions of presently concrescing occasions, which themselves become data for future occasions. Each occasion is a separate time-space region that fuses with the existing time-space regions of the settled past. Instead of conceiving of the leading edge of reality as a moving knife-edge of time, each occasion advances the boundary surface with all other occasions completing their concrescence simultaneous with that occasion. This means that from the point of view of each concrescing occasion, there is a fixed past and an open future. Time-space develops as the potential becomes actual.

In accordance with special relativity, the structure of the various perspectives of actual occasions in the extensive continuum is identified with the kinematic structure of inertial observers as constrained by the light cone. Each actual occasion prehends only the actual occasions in its past light cone and is prehended by those actual occasions in its future light cone.[11] The potential of the past is objectified in the present; that present then becomes an object for future occasions (also with their associated potential), and contemporaries that are outside of the light cone remain unavailable for prehension since they are causally independent.

Substituting 'actual events' with 'actual occasions' here without any significant difference for this purpose, Stapp sees grounds for the unification of relativity theory and quantum theory in the overall framework of a non-linear growth of time-space from basic events.

The relativistic quantum field theory, with its proposal of an open future and fixed past, naturally brings to mind a similar theory proposed in the 1920s by C. D. Broad, namely the growing block universe, according to which physical reality, at any moment, consists of a space-time block of present and past events but no future events, and the block grows with the passage of time. Broad developed this theory to preserve the notion of the passage and the distinct arrow of time against those theorists who viewed the universe to be a fixed block where nothing really happens. Broad, in fact, says he tried to answer 'disturbing arguments' by which J. M. E. McTaggart claimed to prove the unreality of time (Broad 1923: 4). Broad rejected the block universe view in favour of the growing block universe because, like Whitehead, he accepted the notion that temporal passage is not an illusion and becoming is a primary feature of reality. As he expressed the view, he writes:

> When an event, which was present, becomes past, it does not change or lose any of the relations which it had before; it simply acquires in addition new relations which it *could* not have before, because the terms to which it now has these relations were then simply non-entities.
> It will be observed that such a theory as this accepts the reality of the present and the past, but holds that fresh slices of existence have been added to the total history of the world. (ibid.: 66)

Present events are the most recent increments of reality which then become past when fresh events replace them as present. He later called

this position 'absolute becoming', namely the idea of a continual supersession of what was the latest event by a new one. Analogies between
space and time break down when it comes to supersession. This, Broad
argues, is the 'rock-bottom peculiarity of time, distinguishing *temporal
sequence* from all other instances of one-dimensional order, such as
that of points on a line, numbers in order of magnitude, and so on'
(1959: 766).

It is for this reason that Broad, Whitehead and Stapp reject the
Minkowski four-dimensional, block universe in which all events – past,
present and future – exist timelessly. But if this was one of the most
important pieces of evidence in support of the event ontology, then its
modification in the actual-events ontology or relativistic quantum field
theory would seem to undermine the event ontology. That is, having
rejected the space-time block view of the special and general theories of
relativity for the sake of a potential unification with quantum theory,
are we backtracking on our argument? None of these modifications
would imply a return to a substance view or a three-dimensional view of
the universe because the key features of relativity theory are preserved
in this unification, for example the rejection of absolute space, time and
simultaneity, and the elimination of the distinction between space-time
and its physical contents. Moreover, while the Minkowski interpretation of Einstein was certainly influential in Whitehead's development of
his early event ontology, it was not the only influence as he developed
his theory in light of electromagnetism and the early quantum theory.

PRESENTISM, ETERNALISM AND THE GROWING BLOCK UNIVERSE

Philosophers have typically understood the difference between the
three-dimensional and four-dimensional views of reality as two theses
in the metaphysics of time: presentism and eternalism. The former is
generally associated with some sort of substance or object ontology and
the latter with some sort of event ontology. A third position, the theory
of the growing block universe mentioned above, offers an alternative
to both presentism and eternalism. In order to further clarify Stapp's
synthesis and its affinities and contrasts with Whitehead's metaphysics,
it will be helpful to examine briefly these three theories.

Presentists hold that only the present exists and that the present
changes (or moves) since what was present is replaced with a new
present. The central idea advanced by presentists is the notion that
reality is dynamically changing as new events come into existence
and previously present events recede into the past and non-existence.

For the presentist then, neither objects nor past and future times are determinate in the sense of being actual. Or, to put it concretely, my mother and father, computers and mobile phones are real, but my great grandfather, dinosaurs, the dodo bird, my great-grandson, the hovercraft skateboard and human colonies on Mars are not. Aristotle, who appears to hold this position, says in his *Physics*: 'One part of [time] has been and is not, while the other is going to be and is not yet' (1941: 289). Presentists typically affirm McTaggart's A-series, according to which events have the properties of presentness, pastness and futurity. A-series statements are tensed; their truth values change according to the time of their utterance. For example, the statement 'Einstein lives in Princeton' was true when it was uttered in 1955 but false when uttered in 1930 or 1956. But with respect to a tensed statement such as 'Einstein lived in Princeton', when uttered in 1956 or today in 2014, the past has a kind of second-order reality in the sense that statements about such past events have a truth value, but only insofar as the past exists in present memories, fossils or historical records. The future, by contrast, is completely indeterminate and so statements that refer to the future have no truth value.

Eternalists, by contrast, deny both theses affirmed by presentists and instead affirm the view that past, present and future events exist or that all points in time are equally real and the present neither changes nor moves. What is present remains eternally present; past and future are relative to some event or other that is identified by a speaker as a merely indexical 'now'. Eternalists then claim that reality is static since they maintain that what is seen as *our* present does not change nor become past. What happens to be illuminated as *our* present is merely the limitation of consciousness bound by the specious present. In other words, there is no metaphysically or physically privileged present. The indexicals 'now' and 'then' in time are analogous to 'here' and 'there' in space and so there is nothing special about *my* here-now. To claim otherwise would commit a sort of fallacy of temporocentrism, metaphysically adjusted for the perspectives of individuals. Eternalists typically affirm McTaggart's B-series, according to which events have the properties of earlier than, later than and simultaneous with, and they insist that these features do not change. For example, if an event were at one time earlier than another, then it will always be so, irrespective of the time in which it was judged to be the case. B-series statements are tenseless; they are timelessly true or false. According to eternalists, the statement 'Einstein lives in Princeton' is timelessly true because this long event to which this statement refers never ceases to be present in itself. The space-time worm of Einstein's life is eternally fixed in the four-dimensional block, as are the space-time worms of dinosaurs, dodo birds, my great-grandson and the human inhabitants of Mars.

So whereas for presentists 'now' is an *objective present* and the difference between past, present and future is ontological, for eternalists 'now' is a *subjective present* since it is merely the perspective of the present speaker, namely his or her duration of the specious present on eternally fixed events in space-time.

Newton's theory of absolute time, absolute simultaneity and three-dimensional absolute space is perhaps the best example in physics of presentism and the A-series, while the Einstein-Minkowski four-dimensional space-time view is the most obvious example of eternalism and the B-series. McTaggart himself argued in accordance with Einstein-Minkowski when he said that:

> It might be the case that the distinction of positions in time into past, present, and future, is only a constant illusion of our minds, and that the real nature of time contains only the distinctions of the B series – the distinction of earlier and later. In that case we should not perceive time as it really is, though we might be able to think of it as it really is. (1927, II: 11)

Proponents of the third alternative, the growing block universe, claim that when present events cease to be at the leading edge of reality they recede into the past and form the newest layer of the growing four-dimensional block. In making this claim, this theory appears to be a synthesis of presentism and eternalism in that it includes the presentist's claim that the present is real and perpetually moving, and the eternalist's claim that the past is real. This theory also appears to include the presentist's idea of the tensed status of statements. So proponents would agree that a statement, such as 'Einstein lives in Princeton', changes its truth value depending on the time of the utterance, but with one important exception. If uttered in 1930, Einstein is not in Princeton, so the statement is false because there is no reality containing Einstein in Princeton. When uttered in 1955, it was true, and when uttered in 1956 the statement in referring to an event that is now past is still true since the event to which it refers is now fixed eternally in the block. That is, it is not true about the present in 1956, but is true of reality in 1955 for evermore. In agreement with presentists, but in disagreement with eternalists, advocates of the growing block view reject the notion that the future exists and they also reject the idea that statements about future actualities have referents. Broad made the point in contrast to McTaggart, writing:

> You cannot say that a future event is one that succeeds the present; for a present event is defined as one that is succeeded by nothing. We can put the

matter, at choice, in one of two ways. We can either say that, since future events are non-entities, they cannot stand in any relations to anything, and therefore cannot stand in the relation of succession to present events. Or, conversely, we can say that, if future events succeeded present events, they would have the contradictory property of succeeding something that has no successor, and therefore they cannot be real. (1923: 68)

Although it is unclear whether Broad thinks of the fresh slices of existence in the growing block as linear or non-linear, his espousal of the special and general theories of relativity would appear to commit him to the non-linear treatment of becoming in the same manner as relativistic quantum field theory.

Now, with respect to the problem of unification, the main difficulty with presentism is that it is at odds with special relativity. What is present will depend on the frame of reference in which one happens to exist, so there will be no such thing as an absolute or cosmic present for all events. Depending on the speed at which one moves in a given frame of reference, difference sets of events will be simultaneous. The main difficulty with eternalism is the denial by its proponents of the reality of time and change. This presents what appears to be an insurmountable problem for a successful unification with quantum mechanics. Our third alternative, the growing block universe, addresses both of these problems because it is a view of reality in accordance with special relativity and it also treats time and change as real. This theory, however, is not without its own problems. For one, it appears to treat the present and the past as having the same ontological status, the former receding into the block as it is replaced by new present events, but being present and being past are clearly different. This leads to the objection that past events have a strange 'ghostly' state of being once deprived of their status as present on the leading edge of reality. When, for example, Einstein ceases to live in Princeton because this event is now past, what exactly is the status of this event qua past event? It appears that Whitehead and Broad have the same problem, if indeed both espouse some version of the growing block universe. This problem will be addressed in the next section.

All three theories – presentism, eternalism and the growing block universe – are compatible with the idea that time has a direction, that is, asymmetry. Even eternalists, with their Parmenidean insistence that nothing changes, deny the passage of time, but most affirm that there is still an order in which events are directed.[12] Aside from the different philosophies of time, however, some physical laws do not distinguish between the past and the future and therefore support symmetry in nature. For example, in classical mechanics the law $F = ma$

is symmetrical in the sense that it describes the motion of particles or macroscopic bodies without regard to any direction in time. Any physical process described by this law is equally described by a time-reversed process. But now the question is whether all of physics works in this way. Other laws support an arrow of time, such as in thermodynamics in which disorder or entropy increases, in quantum mechanics where we observe the collapse of the wave function and the decay of radioactive elements, and in cosmology where the universe is expanding rather than contracting. Moreover, beyond physics we have much evidence for asymmetry, for example in psychology we remember backwards not forwards, and in causation effects are always preceded by causes and never the reverse. So, the question is: which laws are fundamental or have an overriding status in the depiction of reality? Some physicists, such as Hawking, have argued that the psychological, thermodynamic and cosmological arrows of time support a real direction of time in spite of the fact that many laws of physics do not distinguish between a forward and a backward direction of time (1988: 152). But more fundamentally, if the relativistic quantum field theory outlined above is correct, the asymmetry of time is secured on the basis of the quantum events that create the time-space continuum.

PROBLEMS WITHIN WHITEHEAD'S VIEW

If we adopt the theory articulated above, that is Stapp's modification of Whitehead's metaphysics, it is clear that the past exists as the whole of the space-time manifold. But is this actually Whitehead's own view? One of the most perplexing problems in interpreting his theory of extension and time is the problem of the past. If physical objects are perceived as the immediate past of a concrescing occasion, then the understanding of the ontological status of the past is crucial to sorting out his actual view. Stated quite simply, does Whitehead espouse presentism, namely the view that all of existence is the present becoming of actual occasions and the past exists only as objects of the becoming of present occasions, or does he view physical reality as a growing block universe in which the past exists objectively as past? Eternalism is clearly ruled out by process metaphysics, but there is a dispute among interpreters of Whitehead's metaphysics as to whether he is a proponent of presentism or the growing block view.

When Whitehead modified his earlier theory of a four-dimensional block view to the temporal atomism underlying the process view, he clearly advanced a view in which the future is entirely open from the perspective of the creative activities of the present occasion, yet the theory advanced was meant to be consistent with the special theory of relativity and the denial of a unique cosmic present. Quoting him again on this subject, he says in *Process and Reality*: 'I shall always adopt the relativity view; for one reason, because it seems better to accord with the general philosophical doctrine of relativity which is presupposed in the philosophy of organism' ([1929] 1978: 66). In another passage that clarifies his rejection of absolute time, he writes: 'There is a prevalent misconception that "becoming" involves the notion of a unique serial-ity for its advance into novelty ... In these lectures the term "crea-tive advance" is not to be construed in the sense of a uniquely serial advance' (ibid.: 35).

According to the traditional interpretation of Whitehead's metaphys-ics, we are to take literally his claim that time is a 'perpetual perish-ing' (ibid.: 29, 340).[13] It is necessary for the past to perish in order for novelty to occur in a creative universe. Process, Whitehead says, entails loss. 'The present fact has not the past fact with it in any full immediacy' (ibid.: 340). But 'perishing' does not mean that an occasion that has completed its concrescence ceases to exist as an object; on the contrary, the perishing of subjective immediacy is the point at which the occasion becomes available as an object for the prehension of suc-cessors. Objects, however, exist only insofar as they are objects for the present concrescing subjects. Just as organisms evolve through genetic inheritance or the acquisition of characteristics necessary for their survival, the present concrescing actual occasion is a synthesis of every-thing that came before it. The past does not exist wholly in the present but only in the sense that certain aspects of the past were selected for the subjective aim of the present occasion. In other words, an object is dispersed in the future prehensions as the many become one.

Beginning with Jorge Nobo, the traditional interpretation has been challenged by the idea that Whitehead's real view commits him to the notion that the objects produced in the process of becoming of actual occasions are actual and remain so for the potentiality of every new concrescing subject (1986). According to Whitehead's principle of process, an actual occasion's being is produced by its becoming. *How* an actual occasion becomes constitutes *what* that actual occasion *is* ([1929] 1978: 23, 166). His principle of relativity likewise asserts: 'it belongs to the nature of a "being" that it is a potential for every "becoming"' (ibid.: 22). Nobo argues that commentators have failed to appreciate the true meaning of these doctrines, for Whitehead holds that entities are actual even if they are no longer subjectively immedi-ate. Thus, as Whitehead accounts for the solidarity of the universe, he writes: 'To be actual must mean that all actual things are alike objects, enjoying objective immortality in fashioning creative actions; and that all actual things are subjects, each prehending the universe from which it arises' (ibid.: 56–7). Such passages in *Process and Reality* lend support for Nobo's interpretation. Moreover, Nobo contends that Whitehead's view is a cumulative theory of actuality. When a region of the extensive continuum becomes determinate, it remains so. In this way it can function in later occasions where the data are projected into successors. The solidarity of the universe is the reality of the past.

John Lango suggests a similar revisionist argument by focusing attention on Whitehead's philosophy of time in the context of relativity theory. He argues that neither presentism nor eternalism capture Whitehead's distinctive position because presentism is incompatible with Whitehead's affirmation of the theory of special relativity and eternalism is incompatible with his affirmation of the reality of time. He therefore proposes to see Whitehead's view as a version of the growing block universe. Such a view, as we have seen with Broad, affirms the real existence of the past. Lango calls this view 'past-and-presentism' (2006: 151). The past and present exist, but not the future. Past, present and future, however, are always defined relative to some actual occasion or other.

Lango argues that the growing block interpretation best fits with Whitehead's view because it replaces the unique serial order of creative advance with the notion of durations (sets of actual occasions in unison of becoming) that are not uniquely serial. Moreover, he argues that we would have no way of understanding Whitehead's concept of a nexus if past objects are not actual. What is a nexus if the past does not exist in some sense? The reality or actuality of the past provides a way of understanding how an actual occasion temporalises extension by occupying a basic region that is mediately connected to the basic regions of temporally earlier actual occasions. Understanding Whitehead's theory of extension requires seeing how topological properties of temporal order and metaphysical properties of extensive connection are coherently interwoven.

It is not at all clear which theory of time – presentism or past-and-presentism – is Whitehead's view since the text appears to support both interpretations. And if it is presentism, it would have to be relativised to fit Whitehead's claim that his theory is in accordance with relativity theory, but this no longer looks like presentism. Lango contends that we must focus on what he meant rather than what he said, and therefore Lango argues that the latter view is the only one that makes consistent sense (ibid.: 137). Given the importance of relativity theory as one of the scientific foundations of Whitehead's metaphysics, and its importance to his goal of unifying fragmentary physical theories, I would have to agree with Lango while at the same time pointing out that there is an inconsistency with respect to other parts of Whitehead's system. The inconsistency is substantive in that Whitehead is retaining elements of his earlier four-dimensional view while advancing doctrines of process and becoming in his later view. This is revealed in the debate on the status of the past where advocates of the traditional interpretations focus primarily on Whitehead's theory of epochal becoming and advocates of the revisionist interpretations focus primarily on Whitehead's theory of extension or the theory of relativity. The revisionist interpretations reveal the complexity of Whitehead's system that has been overlooked by the traditional interpretations. Whatever textual evidence can be offered to support the traditional interpretations, it seems clear, given the argument above, that we must accept some variation of the revisionist view as the one that offers some path to a successful unification. While this seems to be the more plausible view, it is necessary to point out some of the problems for the revisionist interpretations. These reveal a deep inconsistency in Whitehead's system.

First, when Whitehead explained time or the past as 'perpetually perishing', he appears to be speaking literally; that is, he meant it. In an essay first published in 1947 in which he summarises his view, he writes:

> Almost all of *Process and Reality* can be read as an attempt to analyse perishing on the same level as Aristotle's analysis of becoming. The notion of the prehension of the past means that the past is an element in the state beyond, and thus is objectified. That is the whole notion. If you get a general notion of what is meant by perishing, you will have accomplished an apprehension of what you mean by memory and causality, what you mean when you feel that what we are is of infinite importance, because as we perish we are immortal. That is the one key thought around which the whole development of *Process and Reality* is woven. ([1947] 1974: 125)

When an occasion perishes, it is no longer subjectively immediate, but contributes its object or superject and then survives only as an element in subsequent prehensions. The data of the past are then redistributed in present occasions. This presents a bit of a puzzle for understanding the past qua past, but Whitehead has another answer, namely the role of God in his system. This is what he means by an occasion's perishing yet remaining immortal in the passage above. God turns out to be a sort of retainer of the whole of cosmic history since only bits and pieces of the past survive directly in the present. This has the rather odd consequence that all historical statements must ultimately be understood as theological since the referents of such historical statements are the retention of the past in God.

Second, if the revisionist view is correct, past events remain eternally fixed in the past, but are deprived of their status as being subjectively immediate. In other words, there is no such thing as ceasing to exist even though an event has ceased to be present. But exactly how we are meant to envisage this realm of past events without subjectivity is puzzling at best. Like the geological strata of the earth full of dinosaur bones, clay pots and foundations of ancient ruins, the past must be seen as dead but not quite gone. Or is the past 'ghostly' in the sense that the full forms of the objects remain in place but somehow lifelessly still present? In other words, the growing block interpretation of the past seems to deprive events of their very eventfulness. But according to Whitehead's ontological principle, everything must be somewhere in actuality. Outside of the creative activity of actual occasions, there is nothingness (Whitehead [1929] 1978: 19, 24). This helps us understand more fully the sorts of claims Whitehead makes regarding extension or objectivity as an abstraction from the concreteness of actual occasions. It also sheds light on the necessity of the one divine actual entity which contains the whole past world as objectively immortal. As he puts it: '. . . the actual world is built up of actual occasions; and by the ontological principle whatever things there are in any sense of "existence," are derived by abstraction from actual occasions' (ibid.: 73). The problem here, as we saw above, also occurred to Broad when he recognised that the past and the present in the growing block view do not have the same ontological status. That is, they do not co-exist in the same sense, for the past is what has been superseded and the present is what is now happening (1959: 767). Becoming is the present in a state of coming into existence whereas the past is completed existence. Whitehead went further in identifying present with subjectivity, but it is this very concept that is the source of the difficulty.

In order to clarify this problem, consider again our now well-worn example of Einstein living in Princeton in 1955. According to the presentist, what was once true in 1955 is no longer the case since the past does not exist, but, for the eternalist, it is, always was and always will be true because this event is eternally present in itself. And for proponents of the growing block view it was true in 1955 and remains the case thereafter that Einstein lived in Princeton in 1955, but in 1956 it becomes false that Einstein lived in Princeton in 1956. Yet as part of the block, Einstein is still there in 1955 trying to work out his unified field theory.[14] Again this raises that nagging question: what is the ontological status of this event once it is no longer present? Now Whitehead's view is a bit more complicated and does not quite match the growing block view on all points. For example, Broad says: 'Nothing has happened to the present by becoming past except that fresh slices of existence have been added to the total history of the world. The past is thus as real as the present' (1923: 66). But for Whitehead, in order for the present to become past it must cease to be subjectively immediate in order to become an object for the prehensions of future occasions. Something has happened to the present by becoming past, namely it must lose subjectivity and so the past is not at all real in the same sense as the present. The past has being; the present is becoming and as such it is not yet determinate. So, on the one hand, subjectivity in the process of becoming is crucial to Whitehead's explanation of how process and novelty occur, yet, on the other hand, his notion of the perishing of subjectivity at the end of an occasion's concrescence results in a view of the past that is virtually unintelligible. Neither Nobo nor Lango correctly account for both of these aspects of Whitehead's theory.

Third, if the past is seen to be actual as a state of being, we have an ontological duplication in Whitehead's system which appears rather implausible especially since it is a violation of the principle of identity. This objection applies more directly to Lango's view since Nobo has argued for an interpretation of Whitehead that involves repeatable particulars. In the 'Final Interpretation' of *Process and Reality*, Whitehead makes it clear that the past is objectively immortal in God's consequent nature ([1929] 1978: 340, 351).[15] That is, while the past is never wholly contained in the present, the past is fully prehended in the divine being, not merely as a copy of what was actual but rather as complete determinate being. The whole space-time worm of Einstein's life, for example, is fixed eternally in the growing block and exists eternally as an object in God's consciousness. But the past cannot exist in two places at once – in the actual past as determinate being and in the being of the divine actual entity.[16]

TIME AND UNIFICATION

The problem of time brings to mind Whitehead's conclusion to his chapter 'Time' in *The Concept of Nature* when he said with typical candour: 'It is impossible to meditate on time and the mystery of the creative passage of nature without an overwhelming emotion at the limitations of human intelligence' (1920: 73). As he sought to develop his mature position in *Process and Reality*, the task of constructing a coherent theory of time remained a central aim of his metaphysics, but, as we have seen, not without considerable difficulty. As many physicists

have noted, only certain parts of Whitehead's metaphysics are compatible with current physics. If we simply focus on time, as I have done in
this chapter, two distinct and inconsistent views of time in relativity
theory and quantum mechanics can be modified to produce one consistent view in which the universe evolves from the epochal becoming
of the basic events. But the ontological status of the past remains a
conundrum for Whitehead. In this regard, his theory requires modification if his view is to become part of any such attempt at a successful
unified theory. Stapp's modification of Whitehead's view takes on
certain aspects of actual occasions that he believes illuminate quantum
phenomena in developing his actual events ontology. These include:
the explanation for the emergence of the discrete from the continuous
and how new events are generated from those that have already been
created. His synthesis, however, appears to avoid the problem of the
past in that the relativistic quantum field theory, like Broad's theory,
does not make any ontological distinction between the actual events at
the leading edge and those past events that form the four-dimensional
growing block. Stapp simply thinks of the present as adding to the
existing space-time framework one quantum jump after another but all
over the advancing boundary surface and without anything perishing in
this process (2011: 90–6).

No one at present would be so bold or foolish to claim to have *the
solution* to the unification of physics, not even those who are currently
working on the cutting edge of theoretical physics. The theory presented
here is no exception, humbly offered for consideration. Whitehead's
theory more closely resembles an outline without the detailed physics
required for the complete, unified theory. This could only be expected
from a theory that was created in 1928 and without foreknowledge
of the physics known today. Stapp claims that an ontological interpretation of the actual events ontology modified by Tomonaga and
Schwinger to create relativistic quantum field theory is the crucial
development beyond Whitehead's metaphysics because it provides the
mathematical structure necessary to understand how a process view of
quantum jumps can be reconciled with the special theory of relativity
(ibid.: 88). Whether Stapp's modification of Whitehead's view will
actually achieve a successful unification of quantum mechanics and
relativity theory remains to be seen. As a contender for the unification,
that is, the Big Problem, it is still very much in the realm of speculation,
but then again so are all the alternative theories.

7. Philosophical Implications

Original ideas are exceedingly rare and the most that
philosophers have done in the course of time is to erect a new
combination of them.
(George Sarton)

A Theory of Everything sought in theoretical physics would unify all the known physical forces and predict the outcome of any experiment that could be carried out in principle. Systematic metaphysicians, such as Plato, Aristotle, Descartes, Spinoza, Leibniz, Hegel, Whitehead and Russell, have also sought 'theories of everything' in terms of a general ontology. These philosophers have attempted to provide comprehensive explanations beyond physics; for even if physics is the basic science, it does not explain everything. Whitehead, for example, produced his philosophy of physics in an attempt to unify the physical sciences, but in his later metaphysics articulated in *Process and Reality* he attempted to go beyond this project with a more general theory that truly sought to explain *everything* – aesthetics, theology, psychology, mathematics, physics, biology, chemistry, cosmology and so on – by a set of interconnected first principles. Accordingly, in this final chapter, I consider some of the most important philosophical implications of an event ontology beyond physics.

As noted in Chapter 1, an event ontology has been proposed by various philosophers as a solution to certain philosophical problems. Here I explore briefly its implications for the mind-body problem, perception and causation, free will, personal identity and moral agency. My intention is to show how an event ontology provides a consistent framework for addressing these problems.

THE MIND-BODY PROBLEM

Most philosophers see the mind-body problem as insurmountable due to the manner in which Descartes conceived of mind and body as two separate substances – *res cogitans* and *res extensa* – with no properties in common. Descartes' contemporaries and his immediate rationalist successors – Princess Elisabeth, Spinoza and Leibniz – recognised the impossibility of explaining the causal interaction between an immaterial, un-extended substance and a material, extended substance. But we are no closer to a scientific nor a philosophical solution today than philosophers were in 1649 when Descartes published his *Passions of the Soul*. The incoherence of substance dualism therefore led to monism, of which idealism explains the fundamental and irreducible reality of mind and materialism explains the fundamental and irreducible reality of matter. Materialism, of course, is currently philosophic and scientific orthodoxy, for which the problem of consciousness remains the sticking point. So, the mind-body problem in the substance tradition has created what Arthur Schopenhauer called the 'world-knot', arguably *the* problem that has bedevilled philosophy. More recent philosophers have characterised it as the 'hard problem', one that might very well be intractable because human beings simply do not have the intelligence nor the cognitive tools designed for understanding the relation between mind (or consciousness) and body (McGinn 1999: 47). In contrast to the defeatist attitude of the 'new mysterians', the event ontology offers a path that is little explored mainly because its proponents reject the one fundamental assumption of most philosophy of mind – that the very concept of substance is an undeniable starting point just as it was for Descartes. It is rather the main obstacle. In a move reminiscent of Berkeley, they have argued that it is physical substance that we do not understand. This being the case, any suggestion that the nature or essence of mind (or consciousness) is something that awaits the results of neurophysiological research or advances in artificial intelligence is fundamentally misguided. The reality of consciousness presents a terrifying prospect for the final truth of materialism. As William James realised, consciousness, however little, is an 'illegitimate birth' in any view that proposes to explain all the facts by continuous evolution from matter (1891: 149).

James, Whitehead and Russell each argued for some version of neutral monism. What all have in common is the conviction that mind and matter are reducible to a more fundamental ontological category. In Russell's *The Analysis of Mind* he surveys the situation in 1921 by observing that psychology under the influence of behaviourism has affirmed a materialistic view while physics post-Einstein has moved away from matter in favour of events, yet if the behaviourists are supposed to view physics as the fundamental science they too should not assume the existence of matter. Russell proposes to follow James by viewing the 'stuff' of the world as neither mental nor material, but rather as a neutral stuff out of which both mind and matter are constructed like a common ancestor (1921: 5–6, 10–11). But what exactly is the 'stuff' of neutral monism? It certainly does not appear to be any sort of Aristotelian nor Cartesian substance that is advanced by any of its proponents.

For James, the 'stuff' is 'pure experience', a very rudimentary sentience at the base of things. This might seem to suggest that it has an interior – that there is something it is like to be this stuff – or subjective features, but this makes experience out to be more mind-like and not at all neutral. It must be neither mental nor physical to be neutral, out of which both mind and matter emerge. But as James says: 'The instant field of the present is at all times what I call the "pure" experience. It is only virtually or potentially either object or subject as yet. For the time being, it is plain, unqualified actuality, or existence, a simple *that*' (1912: 23). Russell is careful in avoiding the use of the word 'substance' to describe the neutral stuff. Yet neither James nor Russell explicitly identify the 'stuff' with events even though this seems the most plausible interpretation of what is meant by mind and matter emerging from some antecedent stuff that is neither. Russell claims that the data of psychology and the data of physics do not differ in their intrinsic nature, but at this point he identifies sensation as the intersection of mind and matter (1921: 143–4, 297). But this also turns out to be an untenable proposal since sensations are unavoidably mental and therefore fail the test of neutrality. It is only later in his work that the 'transient particulars' of which he speaks are understood to be events.

Whitehead appears to offer a version of neutral monism with his notion of the actual occasion, the common ancestor that figures in the nexus of occasions that forms human consciousness and in societies that make up physical objects, including the brain. This might not qualify as neutral monism in the strict sense because the actual occasion is probably best understood as both mind and matter rather than neither mind nor matter. But I take his view to be very close to James' notion of pure experience and James was most certainly a neutral monist. The important point is that the problem of Cartesian interaction disappears since the substance dualism is replaced by an ontology of one type, namely events. The mechanism by which Whitehead explains the emergence of mind in nature is the process of concrescence. As an actual occasion moves through its process of becoming, it can terminate rather quickly and simply repeat the patterns established by the past actual occasions. In these instances, there is negligible opportunity for creativity. The vast multitude of actual occasions constituting physical objects and the vast regions of 'empty space' are of this sort. Repetition of eternal objects in these occasions provides for stability and the continuity of nature. The laws of nature hold in the social order of these occasions because of this repetition. This is what we perceive as inert matter. But now given the place that actual occasions occupy in the extensive continuum, they can be more sophisticated. They can be the fundamental units constituting biological organisms, animal sentience or the fully-developed consciousness of human beings. Here we have creativity of various degrees, namely the possibility of introducing novelty into the process of nature.

Consciousness is the crown of experience wherein the actual occasions function with an intensity of feeling. These highly specialised occasions experience what Whitehead calls 'contrasts', namely the felt oppositions between what is in the actual world and what is sheer possibility, the difference between what is '*in fact*' and what '*might be*' ([1929] 1978: 267). He explains how the actual occasions that constitute consciousness form a single-line inheritance whereby a successor prehends its immediate predecessor in much the same fashion that James described the succession of durations making up the stream of thought (James 1891: 605–42).[1] The character of consciousness as happening in durations that blend together to form the stream of thought and quantum phenomena both share a fundamental character of the epochal nature of becoming.[2] So, at the base of nature there is the atomic becoming of events that eventually form what Whitehead identifies as nexūs and societies, most of which are spatially extended but some, namely the final synthesis from the organism via the nervous system, form a purely temporal nexus. That special nexus, very rare in the scheme of things, is consciousness.

Since all actualities experience the basic sense of passage, or the persistent inheritance of the immediate past, the development of sense organs is not a necessary condition for experience. Our usual distinctions between the organic and the inorganic, the living and the non-living, humans and animals, fail to have ultimate significance for a metaphysics of this sort. Whitehead, for example, contends that 'organic' and 'inorganic' are terms that merely serve a practical purpose in those endeavours for which such a distinction is relevant ([1929] 1978: 102). Thus, when he chose the term 'the philosophy of organism' to describe his late metaphysics, he seems to have in mind biology as the basic science bridging the gap between physics and psychology.

As most neutral monists seem to fall on the subjective or mental side of the dilemma and fail the test of neutrality, the same objection could be raised against Whitehead's proposed solution. The basic principles of life, evolution and consciousness are present in a generalised way in his metaphysical principles. Although it is unclear how much he knew of the mechanics of genetics and evolution or of the specific chemistry behind life processes, he clearly intended to develop a system that accounts for the general features of the natural world (especially change, evolution and novelty) as well as for the higher-grade phenomena of mind and consciousness. As he said in one of his few surviving letters: 'I am trying to evolve one way of speaking which applies equally to physics, physiology, psychology, and to our aesthetics experiences' (Lowe 1990: 333). Whitehead thus created a psycho-physiological language in his attempt to capture the rich diversity of concrete experience.

Actual occasions have both mental and physical poles. The occasions that comprise physical matter have negligible mental poles, those that comprise consciousness have negligible physical poles, and organic matter appears to have a sort of balance between the two. So, is Whitehead, like Russell, proposing a solution that attributes mental characteristics to the so-called 'neutral entities'? While the terminology he chose to describe the polarity in his actual occasions was unfortunate in that it seemed to suggest a mind-body distinction (or homunculus) in the basic units of reality, the real point he is making is that selectivity in nature has degrees. This sounds like panpsychism and many commentators have read him this way, but Whitehead carefully avoided using the term 'panpsychism' himself.[3] Whitehead's emphasis on process or the creative becoming of nature gives new temporal meaning to the way that we typically think of the relationship between mind and body. Accordingly, the events comprising physical matter do little more than repeat the same eternal objects. But the higher we ascend in consciousness, the more chance there is of introducing novelty. Mentality, in this sense, is identified with creative activity.

As we ponder the materialist view of cosmic evolution, we encounter two 'illegitimate births' in James' sense of the term. The first is the evolution of organic life from mere matter; the second is the appearance of consciousness. Both of these developments are relatively recent in the fourteen billion years since the Big Bang.[4] This is essentially the mind-body problem writ large and spread across time on a cosmic scale. The view that makes events the basic entities is more plausible because it permits the emergence of both mind and matter over the course of cosmic evolution, rather than starting out with mere inert matter or the omnipresence of consciousness and then trying to explain how one or the other appeared, how one is reduced to the other or is entirely eliminated.

PERCEPTION AND CAUSATION

Perception has been a problem for substance philosophy at least since the time when the Scholastics tried to make sense of what was being received by the knowing mind, given Aristotle's hylomorphism. In his 'New way of ideas', Locke substituted representational realism for direct realism but, for many philosophers, including Whitehead, this view led to scepticism instead of a solution to the original problem. Having rejected the metaphysical substratum view, and the bifurcation of nature implied by this view, Whitehead revises the very meaning of the subject-object relation in perception. The 'subject' is the present concresing occasion that prehends the object, which is the finished concrescence of the multitude of occasions settled in the immediate past. So, the creative activity of nature is conceived as a process of subject becoming object as the determinate past flows into the subjectivity of the present, which upon completion or 'satisfaction' itself becomes an object for the future. Instead of viewing perception of an object as contemporaneous with the perceiving subject, any perception takes time. Perception is always perception of a past world. The crucial point for an event-based view of perception is the immanence of the past in the present perceiving subject.

Since Whitehead's event ontology radically revises the perceptual process, this has a direct consequence for our understanding of causation. In fact, perception and causation are intimately connected in his view. Nowhere is this brought out more clearly than in his refutation of David Hume's analysis of causation.

The central problem of causation is to explain how causes can be genuinely efficacious, as having come about through the compulsion of the past and resulting in some future effect. But this is exactly what Hume cannot find in experience. As he explains:

> It must certainly be allowed, that nature has kept us at a great distance from all her secrets, and has afforded us only the knowledge of a few superficial qualities of objects; while she conceals from us those powers and principles on which the influence of those objects entirely depends. ([1748] 1927: 32–3)

Since Hume's attack on the purported necessary connection in causal explanations, philosophers have been wary of discussing events as causally efficacious. For such thinkers, Hume destroyed a relic of anthropomorphism intimately tied with our explanations of natural phenomena. Instead of viewing causation as a propensity in events to produce effects, his analysis has focused on causation as an external relation between separate events. Dorothy Emmet in *The Effectiveness of Causes* has made this the basis for her criticism of Humean accounts where all consequence is understood as mere sequence. Where events are simply described in terms of different properties at different times, we invariably have what she calls a 'Zeno Universe'. As in the paradox of the moving arrow where 'motion is defined as the occupation by one entity of a continuous series of places at a continuous series of times', movements and transitions get lost (1985: 10). Events are simply located in distinct spatio-temporal regions but there is no sense of the activity in one stage carrying over into the next.

This is exactly Whitehead's main disagreement with Hume. He argued that Hume's analysis of causation rests on a faulty theory of perception and so he endeavours to show how the analysis of perception, in terms of individual impressions of sensation, contrasts strikingly with the way that we actually experience the world. Like Hume, Whitehead contends that experience is atomic, but, unlike Hume, he maintained that continuity arises from the epochal nature of the atoms of experience.

Hume's theory of impressions lies at the heart of his empiricism. Each impression is distinct and separate; in Cartesian terms, each requires nothing but itself in order to exist. As the maxim is stated clearly in the *Treatise*, Hume writes:

> Again, every thing, which is different, is distinguishable, and every thing which is distinguishable, is separable by the imagination . . . My conclusion . . . is, that since all our perceptions are different from each other, and from every thing else in the universe, they are also distinct and separable, and may be consider'd as separately existent, and may exist separately, and have no need of anything else to support their existence. ([1739] 1958: 233)

But when Hume turns his attention to the external world to consider the operation of causes, his inability to discover any power or necessary connection that binds the effect to the cause results from the atomistic framework of the theory of impressions. Just as any one impression admits no relevance to its predecessors or successors, any one event in the external world admits no necessary connections. For Hume, the only thing we can know about causation is that one event is temporally prior to another, that the two events are contiguous in time and space, and that there is a constant conjunction between what is thought to be the cause and what is thought to be the effect. The so-called 'necessity' then is nothing but the determination of the mind to pass from causes to effects, and from effects to causes according to their repeated union. But it does not arise from experienced facts as is commonly supposed.

Whitehead, however, argues that if we attend to our experience carefully, we will find that Hume's analysis addresses one part of our perceptual experience, but what Hume takes to be fundamental in experience is a result of our highly-refined sense organs. Whitehead therefore distinguishes two distinct perceptive modes that he calls 'presentational immediacy' and 'causal efficacy' (1927: 31, 1978: 176). Presentational immediacy corresponds roughly to Hume's impressions of sensation. It is our clear, distinct consciousness of the external relations in the contemporary world. The data it presents are sharp, clear-cut presentations of sensa, spatially located and temporally self-contained. Presentational immediacy, however, does not supply any information as to the continuity of events, that is, the relevance of the past or the anticipation of the future. This led Whitehead to adopt a more primitive and fundamental mode of perception as causal efficacy. Presentational immediacy is an elaboration upon certain aspects of what is already present in the flow of experience and in essence involves a projection of images onto the present. Where evolution has given us such acute receptors, we naturally focus our attention on what is clear and distinct in experience. With our highly-developed sense organs, we grasp the details of passage in the immediate perceptual field and identify them as sensa, that is various patches of colour, size, shape and texture. But this mode of perceptual activity presupposes the whole context of passage in time. Causal efficacy is therefore the mode of perception that is simply felt as an awareness of the persistent inheritance of the given past world. It is our experience of the immediate past in the process of becoming present. The feelings it transmits are vague, but as Whitehead puts it, 'their explicit dominance in experience may be heightened in the absence of sensa' ([1929] 1978: 176).

By explaining something rudimentary in immediate experience that connects the past with the present, Whitehead contends that this is the only way of answering Hume that will yield a basis for causal relations.[5] Although causal efficacy is a relatively crude form of perception that does not produce any distinct sensation for our consciousness to fix upon, it nevertheless presents itself as the fundamental feeling of existence. Our dim consciousness at half-sleep, the thumping of our heart beat and various types of visceral feelings all suggest continuous becoming in the mode of causal efficacy.

The difference in world-views between Whitehead and Hume, with respect to the primacy of causal efficacy or presentational immediacy, was captured succinctly in Russell's recollection of an argument with Whitehead. In speaking of his own philosophy, Russell wrote:

> It was Whitehead who was the serpent in this paradise of Mediterranean clarity. He said to me once: 'You think the world is what it looks like in fine weather at noon day; I think it is what it seems like in the early morning when one first wakes from deep sleep.' I thought this remark horrid, but could not see how to prove that my bias was any better than his. (1963: 39–40)

For Whitehead, whenever clear-cut perceptions of presentational immediacy are minimised, such as the experience of waking from deep sleep, the rudimentary basis of experience reveals itself as an essentially dumb throb of existence. Acute visual perceptions are appended to this experience once consciousness comes more clearly into focus. So for Hume and Russell, the paradigm for perceptual experience seems to be visual sensation, whereas, for Whitehead, the broader involuntary sensations that are revealed through bodily organs are primary. The former provides us with a sense of sharpness and clarity, while the latter, a type of kinaesthetic experience, provides us with a rudimentary sense of inheritance and continuity.

Whitehead therefore argues that philosophers, in their analysis of perception, have ignored perception in the mode of causal efficacy by concentrating on those distinct impressions that are mainly revealed through our visual perceptions. The obvious result of analysing perception solely in terms of presentational immediacy is that it is difficult, if not impossible, to explain causal relations.[6] As he makes the point, 'Hume's polemic respecting causation is, in fact, one prolonged, convincing argument that pure presentational immediacy does not disclose any causal influence' ([1929] 1978: 123). The conclusion of any such analysis in the mode of presentational immediacy is that events in the contemporary world are causally independent of one another, but this is because Hume has inverted the relationship by making the experience of causal efficacy dependent upon presentational immediacy. Presentational immediacy unveils the world at an instant and, when taken alone, will inevitably fail to reveal any intrinsic connection with the past or future. This is Hume's conclusion. But by neglecting the whole antecedent world that grows continuously into the present, Hume missed the concrete sequence of causal relatedness. Together, presentational immediacy and causal efficacy constitute the fully alert mode of human perception as 'symbolic reference', which includes an integration of our awareness of the temporal process of the world and the definite spatial sensa discerned in the immediate present.

One particular flaw in Hume's analysis of perception and causal relations is to be found in the type of examples he uses to make his point. In fact, his most common example of billiard balls striking one another on a distinct boundary of a billiard table typifies the world of presentational immediacy (1927: 29, 63, 1958: 164). Each ball is a hard and precise entity distinctly located in a spatial region. The trouble, however, is that our experience of causal relations is not always so clear cut, especially when we are attempting to discern the dynamics of events. In many cases our perceptual faculties fail to identify exact points of contact. Whitehead thus provides a counter-example in which a simple reflex action shows the inadequacy of Hume's analysis. In the dark, an electric light is suddenly turned on and a man's eyes blink. The sequence of perceptions, in the mode of presentational immediacy, is as follows: (1) the flash of light, (2) the feeling of eye closure, and (3) the instant of darkness. Although the flash maintains its priority over the other two perceptions, the last two are indistinguishable as to priority. According to Hume, there is no perception of the flash *making* the man blink. In his analysis, there are merely two perceptions – the flash and the blink. Thus in his interpretation of the man's experience, the man is simply in the habit of blinking after flashes of light. One event simply follows after the other without the latter being the consequence of the former. But in Whitehead's view there was a perception in the mode of causal efficacy, namely the *feeling* of causality and of the compulsion to blink. The man 'feels that the experiences of the eye in the matter of the flash are causal of the blink' and it is this feeling of causality that is precisely what enables the man to distinguish the priority of the flash ([1929] 1978: 175). Even Hume would have to admit that the impression of the flash of light had a certain force and vivacity, thereby suggesting the force of the immediate past upon the present. Although Hume has correctly accounted for the events in the mode of presentational immediacy, he failed to account for the manner in which the present results from the immediate past. The additional perception added by causal efficacy is the feeling of cause, which in this case is the force of the light on the eye.

Hume may very well counter at this point by arguing that the feeling of causality is nothing other than the flash, and that this cannot show that there is a *connection* between the eye closure and the instant of darkness, only a *conjunction*. As he puts the point clearly:

> What alteration has happened to give rise to this new idea of *connection*? Nothing but that [the man] now *feels* these events to be *connected* in his imagination, and can readily foretell the existence of one from the appearance of the other. When we say, therefore, that one object is connected with

another, we mean only that they have acquired a connection in our thought and gave rise to this inference by which they become proofs of each other's existence. ([1748] 1927: 75–6)

What is perhaps the most significant thing to notice here is that Hume does consider the possibility that the idea of necessary connection is one that originates upon reflection on a *feeling*. But note again that the feeling from which this *idea* arises is expressed solely in terms of presentational immediacy. Hume does not even allow for the possibility that the cause and the flash are one and the same, that is the feeling of *connection* is not grounded in the compulsion to blink but rather is viewed as a product of the imagination. In his view, the so-called 'necessity' is nothing other than an attitude that arises in the man's mind when he is confronted with the repeated relationship of the flash and the subsequent blink.

By contrast, the antecedent flash, in Whitehead's view, is not a self-contained item, even though it retains a certain distinctness in presentational immediacy. The feeling of causality is a much more general type of perceptual activity than is captured by Hume's concept of impressions. As a cause, the flash is a particularly forceful element in the ongoing flow of events that pass into the man's experience.[7] From the perspective of the man experiencing a perception in the mode of causal efficacy, the *instant* of the flash is specious in the sense that it is never just here-now but rather spread over a duration in which the past is merging into the present. This is the problem that Whitehead addresses with his 'fallacy of simple location' (1925: 49).

We should note here three important points about Whitehead's analysis of causation. First, in his counter-example, he has placed the perceiver in a context in which he is more than a passive spectator of constant conjunction. The perceiver is acted upon externally; that is, he is affected by an external event, that event being the flash of light (and whoever flipped the switch) that acts as a causal agent. Second, in Whitehead's view of causal efficacy, the external flash becomes an internal constituent in the perceiver's experience. It is exactly this relationship of the immanence of the past in the present that gives causes their effectiveness. And third, given his modified version of perception, the idea of a necessary connection is an inevitable product of experienced facts, which is exactly what Hume denies.[8]

Whitehead argues that the idea of necessary connection remains an enigma for Hume because of his adherence to a subject-predicate form of statement that *reflects* a substance-accident metaphysics. He tacitly treats the 'soul' as subject and 'impressions' as predicate. For example, in speaking of impressions of sensation and reflection, Hume writes:

'the first kind arises in the soul originally, from unknown causes' ([1739] 1958: 7). Although he explicitly repudiates the substance-accident metaphysics, his use of 'soul' or 'mind' functions as a passively receptive substance while 'impressions' happen *seriatim* as a private world of accidents (Whitehead [1929] 1978: 137–8). Such applications elicit the inevitable conclusion that we are unable to account for the relations between impressions.

As a result of Hume's analysis of causation he argued that we can give no justification for our expectation that the future will resemble the past, that is, the problem of induction, since such inferences depend upon the uniformity of nature. But does the Whiteheadian view of causation fare any better? With regard to the uniformity of nature we may ask him the same question raised by Hume: what reason does experience give us for expecting the future to resemble the past? Although Whitehead maintains that predictions about the outcome of future events remain merely probable, he does provide a reasonable ground for such inferences. This essentially involves an interpretation of events as being dependent upon one another via the asymmetry of time. Each event comes into being through an inheritance of its causal past, provides its own measure of effectiveness and conditions future events beyond itself. With this framework in place, the expectation we have for the future is that there will be a general maintenance of order in the environment by the internal relations of events and the fact that these events merely repeat the same patterns of eternal objects. Moreover, concerning the uniformity in nature, Whitehead's view requires only that the immediate past shall be related internally to the immediate present. What we do know about the future is that what *will* happen will be derived in some way from the present and will be subject to those limitations that the present sets for the future. Our envisagement of future events determines our present actions and gives them their significance as causal agents. We plant crops, plan railway timetables and predict eclipses and the appearances of comets. In the analysis of present events, there is simply potential as anticipation of future events. As Whitehead makes this point, he writes:

> Thus, according to the philosophy of organism, inductive reasoning gains its validity by reason of a suppressed premise. This tacit presupposition is that the particular future which is the logical subject of the judgment, inductively justified, shall include actualities which have close analogy to some contemporary subject enjoying assigned experience; for example, an analogy to the judging subject in question, or to some sort of actuality presupposed as in the actual world which is the logical subject of the inductive judgment. It is also presumed that his future is derived from the present by a continuity of inheritance in which this condition is maintained. There is thus the

presupposition of the maintenance of the general social environment – *either* by reference to judging subjects, *or* by more direct reference to the preservation of the general type of material world requisite for the presupposed character of one or more of the logical subjects of the proposition. (ibid.: 204)

Recognising the importance of causal efficacy in experience and 'the continuity of inheritance', we have a reason for putting a *limited* trust in induction. The general patterns of our environment in which certain events take place provide a reasonable basis for making predictions and these patterns constitute the essential foundation for our ability to formulate laws of nature. However, Whitehead maintains that these laws of nature are entirely dependent on the individual character of the things constituting nature. Thus as we noted in Chapter 5 above, we must conceive of the laws themselves as evolving concurrently with the things constituting the environment.

The search for causes in nature remains as elusive as ever. Hume's argument is a powerful demonstration that science is endlessly engaged in a process of conjecture and refutation and never arrives at truth. But Whitehead's argument for causation strongly suggests that the direct evidence for connectedness (or the lack thereof) is highly dependent upon the manner in which perceptual experience is analysed. Hume's argument depends on a presupposition of psychological atomism at the centre of his view of perception, a fact of which he seems unaware. If Whitehead is right, however, the basic experience in which presentational immediacy occurs does in fact reveal connections in experience rather than mere conjunctions. As long as one remains at the level of clear and distinct qualities and spatial relations discerned in presentational immediacy, one remains on a fairly abstract level of thought where isolated objects are viewed in purely external relations. Something crucial, however, is added by the experience of inheritance.

FREE WILL

Are we free or determined? Determinism is the thesis that there is at any instant exactly one physically possible future. The free will thesis, by contrast, proposes that agents have the power to create and sustain their own ends and thereby denies that there is one and only one physically possible future. The unmodified four-dimensionalism of the special and general theories of relativity, according to which all events – past, present and future – are ontologically fixed within space-time, is one version of determinism. And as we saw in Chapter 6, with McTaggart's B-series, all events are ordered as 'earlier than', 'later than' or 'simultaneous with' one another, but there is no sense in which the present of a speaker has any kind of ontological status different from those events identified as 'past' or 'future' of that speaker. This position is sometimes called *ontological determinationism*. According to this theory, the distinction between the past and the future is on a par with 'left' and 'right' in spatial position. We can no more alter the future of our present moment than we can alter the past. In other words, there is absolutely no wiggle room for free will. In addition to Einstein and Minkowski, determinationism has been held by many monistic philosophers, such as Parmenides, Zeno, Spinoza, Bradley, Royce, Sprigge and others of various philosophical stripes, such as Santayana, McTaggart, Gödel and Quine. In the other more familiar deterministic view, *causal determinism* or *epistemological determinism*, the future is determined and predictable, given sufficient knowledge of past events and the relevant laws of nature. The causal determinist believes that future events will be causally determined by present events and are, therefore, inevitable, even though they do not as yet exist. This view is often attributed to Newton and the advocates of mechanistic materialism. An intellect that is powerful enough could, in principle, predict every event in the history of the universe given knowledge of Newton's laws and the current position and momentum of each particle.

While the event thesis is neutral with regard to the free will versus determinism debate, Whitehead's particular version of the event ontology in *Process and Reality* seeks to secure free will in all actualities in nature. Our experience of freedom is not an illusion; it is grounded in the fundamental reality of becoming in the universe – the self-causation of actual occasions ([1929] 1978: 88). Much of Whitehead's metaphysics, as we have seen earlier, aims to explain just exactly how novelty is possible. The critical point concerns his understanding of what it means to be present. That is, an actual occasion that is present is subjectively immediate, and to be subjectively immediate is to be indeterminate. To become a publically available datum for future occasions, the present concrescing occasion has to complete its process of becoming or reach a satisfaction. So, on this understanding of what it means to be present, that is not determinate yet, it follows that the future from the perspective of that occasion is genuinely open. The present itself is a transition in which potentiality is in a process of actualisation. This is what provides us with a real sense of choice regarding our present actions, and the reason is that we really are making a choice among many possible futures. But we also experience limitations that are established by what has been determined in the past.

As discussed in Chapter 6, the determinism of the theory of relativity had to be modified in order to accommodate quantum mechanics and the creative advance of nature in a theory based on process. It was, in fact, the very indeterminism of quantum mechanics that undermined the stronghold of determinism in physics. Ontological determinationism is therefore rejected, but there is a sense in which causal determinism, as described in classical physics, is explained by the repetition of the past. As discussed above, the actual occasions that comprise the greater part of nature terminate early in the concrescent process and thereby provide stability in physical matter and in the forces of nature. What will be is causally determined by the past; the future is thus predictable on this basis. It is here that the periodicity of nature is established and the basic laws of mechanics operate with a uniform precision that, for example, can be achieved in the calculation of the motion of the tides and the orbits of the planets. While the microscopic and macroscopic entities behave differently, it should be clear that at the very base of nature any actual occasion is an active process of self-creation. The issue of freedom is a matter of degree depending on the sophistication of the occasion and its place in the extensive continuum. On this score Whitehead turns out to be a compatibilist on the free will/ determinism debate.

Throughout this work, we have seen that Whitehead defends our experience of the passage and direction of time and our experience of causation. The situation with free will is not significantly different. Any system of philosophy or science that bifurcates nature in order to make a distinction between reality and mere appearance must then explain our experience as falling under the latter category. Ontological determinationism, which results from the view of the universe as a fixed block, is one such example of this bifurcation.

PERSONAL IDENTITY

The problem of the identity of the self or person, or, more generally, the problem of sameness throughout change, has had a long history in substance philosophy, beginning with Aristotle's notion of the soul as the core essence of a living being. What makes us the selfsame entities throughout changes in time is the identity of the substance. Rationalists such as Descartes have identified the substance to be the indubitable *cogito* and Leibniz followed suit with the idea of the monad. Empiricists such as Locke and Hume, however, raised doubts about the experience of substance that supposedly served as the foundation of the self. And while Locke nonetheless continued to believe in substance as a substratum underlying experience, it was his famous memory criterion, 'as far as consciousness can be extended backwards to any past action or thought, so far reaches the identity of that person' that answered the question ([1748] 1927: 247). Identity is preserved not because the different states of consciousness occur in one and the same substance, but rather because a later state 'can be extended backwards' to a former state. Hume characteristically remained sceptical of how the theatre of impressions produced any idea of the same self. Resemblance among the succession of impressions gives rise to the idea of identity, but it is the closeness of the related objects in experience that is confused for identity ([1739] 1958: 254). Hume is often recognised as holding an event view of personal identity, even though he clearly did not embrace an event view as fundamental ontology.

Some philosophers, most notably Strawson, have maintained that without substance the whole bottom drops out of our world and an exclusive event ontology would not only undermine the possibility of personal identity, it would also undercut the ability to re-identify ordinary objects. Plato, in the *Theaetetus*, comes close to defending a position of this sort when he rejects Heraclitus', Protagoras' and Empedocles' view that all things are the offspring of flux and motion. As Plato has Socrates explain to the young mathematician, Theaetetus:

> I will explain, and tell you of a high argument, which proclaims that nothing in this world is by itself *one*, or can rightly be called *this* or *of this kind*; but if anything is termed great, it will appear to be also small, if heavy, light, and so forth. There is no *one thing*, no *this*, and no *such*. It is from motion, and change, and admixture with each other, that there *come to be* all those things which we declare to *be*, speaking incorrectly, for there is no being at all, but only perpetual becoming. (1953: 247)

While it might be the case that strict identity is denied by the event ontology (characterised by Plato here as 'perpetual becoming' in the lovers of flux), the pragmatic notion of an object or the identity of the person is retained. For Whitehead, each event in a sequence or nexus is a unique particular and the nexus has a derivative ontological status. So, identity in the strict sense is to be found in the event as the particular, not in the nexus, which for him is the change from event to event.

At one point in *Process and Reality*, Whitehead speculates: 'The final percipient route of occasions is perhaps some thread of happenings wandering in the "empty" space amid the interstices of the brain' ([1929] 1978: 339). This is what he means by the nexus of events (or actual occasions) that constitute consciousness and form the basis for personal identity. Locke's memory criterion is highly relevant to this analysis. The common-sense notion that our identity stretches as far back as our memory extends has a metaphysical basis in Whitehead's notion of prehension. As explained in Chapter 4, prehension is a

relation between the present and the past. The present concresing occasion prehends the immediate past and the many become one in a novel
concrescence. Prehension, conceived in this way, is a metaphysical
generalisation of memory and perception. As the data from the past are
prehended, what is compatible with the subjective aim is included in the
novel entity, and what is not, is excluded. Both perception and memory
are selective. Nothing from the past survives in its entirety, but what is
most valued continues to be repeated in the process of selection. Our
personal identity, like the identity of any physical object, is a matter of
degree, and it is dependent on the continuity of the eternal objects or
subjective forms that we recognise as that 'thread of happenings'. So, as
a human being develops from zygote to embryo, foetus, infant, adolescent, adult and so on, there is no absolute sameness among difference
but merely the continuity of a certain collection of physical and mental
characteristics that we are satisfied to call the 'same person'. We come
into this life by degrees, personhood accumulates by a gradient process
over time, and at the end of life, unless by abrupt exit, we depart in
degrees.

Whitehead's view of personal identity has been compared to more
recent theories, such as Derek Parfit's defence of the psychological
continuity theory and older theories proposed by Buddhist philosophers.[9] In all these theories, the continuity of consciousness is the basis
for personal identity rather than a substantial self. And here the use of
'identity' is perhaps misleading in the sense that all such theories deny
the strict sense of identity that is associated with the same substance.
The process view also rejects any sort of metaphysical ensoulment typically asserted by followers of a substance view.

MORAL AGENCY

A person is a morally responsible agent. But how can this be so without
the identity of substance that seems to be required for moral responsibility? This presents a challenge for the event theory of personal identity.
Dorothy Emmet, in fact, raises this very problem for an exclusive event
ontology when she claims, in accordance with Donald Davidson, that
an adequate conceptual scheme requires both objects and events. Active
processes need active constituents and moral agency requires persons as
enduring objects of action. In this regard, she says: 'The capacity to act
and be acted on is my general criterion of being a thing as distinct from
being a process or event' (1992: 49). Locke thought it obvious that in
order to be accountable for one's actions one must be able to recognise
them as one's own, so awareness of what one is doing is required, as is

the ability to remember having done it. Emmet appears to agree with Locke when she says that those who hold a view of the self as a succession of states are guilty of a 'no ownership' view that threatens to dissolve moral responsibility. She writes:

> I can be concerned about the consequences of my actions in future states, whosesoever they are, but I cannot acknowledge responsibility for actions as *mine*. If I am to take responsibility for my past actions I shall need to think of myself as enduring, as being wholly present in all stages of my history, and not as perduring, as only partially present in any temporal part ... A pure process could perdure in temporal parts, but it could not have the personal responsibility of an enduring self. (ibid.: 50)

So it seems Emmet's argument can be stated concisely: if the pure event or process ontology is correct, then there would be no basis for moral responsibility, but as there is a basis for moral responsibility, the pure event or process ontology must be incorrect. Whitehead's response to this problem is contained in the following statement when he writes: 'What needs to be explained is not dissociation of personality but unifying control, by reason of which we not only have unified behaviour, which can be observed by others, but also consciousness of a unified experience' ([1929] 1978: 108). In his metaphysical principles, it is the concept of the subjective form that addresses this unifying control and behaviour in the self across time. It is how the subject is feeling the objective datum and carries into the feeling its own history. That is, the concrescing actual occasion feels presently what the past actuality felt when it was actual or subjectively immediate. Consciousness in human beings is a specific intensity of the subjective form. This accounts for the enduring self and the recognition of 'actions as *mine*'. In other words, even though Whitehead certainly espouses an exclusive event ontology, in his case it is doubtful whether this results in a no-ownership view that dissolves moral responsibility. The first premise of Emmet's argument is false and so is the conclusion.

The failure to find a substantial self in experience does not imply that I have no moral responsibility for my actions. All that is required is unifying control in the stream of consciousness. I recognise that my present moment has emerged from my immediate past and I am connected via memory to my past moments. Both are a matter of degree, but the latter is particularly important when it concerns accountability for my actions. I have much in common with the person I was yesterday, five years ago and even fifteen years ago, but very little in common with the person I was forty years ago and almost nothing with the adolescent and child. It is true that there is a core of values, beliefs and abilities. These were acquired slowly and yet they constantly face revision by the tribunal of experience. Nothing here serves as a disproof of the event ontology.

It is well known that Whitehead did not produce a moral philosophy, but this does not mean his philosophy is amoral. This omission is an anomaly for a systematic philosopher, due mainly to the fact that he explained value as part of his general metaphysics. In Whitehead's system, morals are derived from axiological-metaphysical principles. There are no Categorical Imperatives nor Principles of Utility; rather the lure for beauty is built into the cutting edge of reality, but this value selection also involves the balance of ideal opposites. Where harmony becomes monotonous, there is the quest for novelty and variety, and wherever novelty and variety become chaotic, there is the renewed establishment of order (ibid.: 339). In this regard, there is every opportunity for the achievement of civilisation guided by the wisdom of sound judgement and for tragedy resulting from the destructive revolts of self-interest. Many have seen in Whitehead's system a non-anthropocentric ethics in which the omnipresence of sentience and intrinsic value in nature become the basis for moral principles.

CONCLUSION

As I have argued in the preceding chapters, an event ontology has advantages over the traditional substance theory insofar as it affords greater empirical adequacy, explanatory power and unification within the context of modern theoretical physics. It also addresses the concern for ontological parsimony and provides a more fruitful approach to the mind-body problem, free will and determinism, personal identity, causation, perception and epistemology, since the main focus of the theory is an ontology of one basic type of particulars. In this way we avoid the incoherence presented by the basic dualisms in substance/property thinking. Davidson's point is well taken – a theory that does not recognise events among the ontological citizenry of nature deprives us of the coherence of our explanations. With the revisionary approach discussed above, however, Davidson's point is taken even further by dispensing with substance altogether in favour of a pure event ontology. As to the concern for the loss of substance, perhaps Berkeley put it best when he said:

> That the things I see with my eyes and touch with my hands do exist, really exist, I make not the least question. The only thing whose existence we deny is that which philosophers call matter or corporeal substance. And in doing of this, there is no damage done to the rest of mankind, who, I dare say, will never miss it. ([1710] 1957: 39)

As with all revisionary metaphysics, everything changes and nothing changes. Our conceptual framework undergoes radical revision, but we carry on in everyday life with our ordinary patterns of speech and perception.

The very idea of alternative conceptual schemes has been met with strenuous resistance by descriptive metaphysicians, such as Davidson, in part because the idea of conceptual relativism is thought to be incoherent or to lead to the abandonment of truth (1984). Science, however, is not a mere alteration nor a 'suburb' of our present conceptual apparatus; it is a revision of it, if indeed a revolutionary idea passes the test of rigorous experimentation and explains a broader range of phenomena. This being the case, adjustments at the more general end of our view of the world are also necessary – hence the open door to revisionary schemes.

Strawson claims that revisionary metaphysics is at the service of descriptive metaphysics because any attempt to produce a better structure of thought about our world already presupposes the point of departure of our homely conceptual apparatus (1959: 9). Whitehead might very well concede this point yet insist that we need not remain in continual service to our ordinary conceptual scheme, since devotion to the methodology of conceptual analysis and the suppression of speculative novelty eventually result in the fatigue of reason ([1929] 1958: 18–21). With an uncanny anticipation of Strawson's famous distinction, Whitehead writes:

> The Fallacy of the Perfect Dictionary divides philosophers into two schools, namely, the 'Critical School' which repudiates speculative philosophy, and the 'Speculative School' which includes it. The critical school confines itself to verbal analysis within the limits of the dictionary. The speculative school appeals to direct insight, and endeavours to indicate its meanings by further appeal to situations which promote such specific insights. It then enlarges the dictionary. ([1938] 1968: 173)

What we gain in clarity and safety by remaining in the confines of our homely conceptual scheme sacrifices the adventure of the speculative urge.

There is a great wealth of ideas in Whitehead's metaphysics that demonstrates his astonishing ability to generalise from the natural sciences and see connections that unify fragmentary notions into one grand conceptual scheme. Some of these ideas will be judged unsatisfactory given the current state of physics, but I think the more relevant question is: which provide fruitful direction for rounding out our general theory of the world? In this book, I have made my case for adopting Whitehead's approach to unification on the basis of two basic ideas – events and process – upon which it seems to me fundamental ontology provides direction at the most general level for working out solutions to the problem of unification. Whether or not Whitehead's ideas play any role in the theory that is finally developed and accepted, it seems likely that: (1) as Einstein predicted, it will be some kind of field theory, (2) either relativity theory or orthodox quantum theory (or both) will need to be modified, (3) the question about the nature of time will be fundamental; either time will be treated as an illusion, a local phenomenon of a multidimensional universe (or multiverse), or it will figure as a key feature of reality, (4) the theory will provide testable predictions, and (5) the theory will exhibit broad explanatory power.

Aside from the philosophically-minded physicists who were influenced by Whitehead, his unified theory has had little impact on the course of theorising in the twentieth and twenty-first centuries in the fields of physics and cosmology. Some physicists, such as Steven Weinberg, have claimed that philosophy provides no useful guidance to science. The insights it presents, he asserts, 'seemed murky and inconsequential compared with the dazzling successes of physics and mathematics' (1994: 168). Physicists, in their search for the final theory, have therefore been more like hounds sniffing about the ground rather than hawks high above seeing general patterns in philosophy.[10] It is not clear, however, that the likes of Whitehead, Russell and Quine were on Weinberg's reading list, the very ones who sought to correct the mistakes of other philosophers by advancing the proper relationship between science and philosophy instead of treating them as separate endeavours. It was rather the arcane, scholastic quibbles of philosophy that failed to impress Weinberg and here there is much with which to agree. The great achievement of his quantum electroweak theory was due more to the work of the hounds but the grand scale of the final theory might need some help from the hawks, that is top-down rather than bottom-up theories.

In response to Weinberg, then, it is not so much that philosophy has nothing to contribute to science, but rather what the right kind of philosophy has to offer, namely, a scientifically-informed metaphysics. The influence of bad philosophy – such as logical positivism or mechanistic materialism – on science is clearly established in history. Other kinds of philosophy, such as linguistic analysis, postmodern relativism and social constructivism, have further contributed to the impression among most scientists that philosophy in general is irrelevant or even subversive to scientific enquiry. There is, however, no escape from philosophy in the same way that there is no escape from some conceptual framework by which we interpret our experience.

The interdisciplinary appeal of Whitehead's metaphysics has spawned interesting developments among physicists, biologists and chemists, but given the dominant tendency of specialisation in the disciplines this is generally treated with suspicion. Even in philosophy, historically the most comprehensive and fundamental of all disciplines, George Lucas rightly notes the driving trend is to 'professionalize philosophy as a distinct discipline, to delineate its generic concerns in contrast to those of other related disciplines, and most of all to exclude from this discipline all who do not share the requisite background, scholarly interests, and methodological commitments' (1989: 131). Overspecialisation of the disciplines has produced a situation in which narrowness is considered a virtue even though it potentially deprives us of the penetrating idea from cross-fertilisation. As Whitehead himself noted, 'each profession makes progress, but it is progress in its own groove', each contemplating its own set of abstractions, but none of which alone is adequate for comprehension of human life or the physical universe (1925: 245). The problem of unification requires a comprehensive vision that extends beyond hyper-specialisation and the right balance of imagination and specialised, technical work.

It is of course possible that there is no *final*, unified theory, but only (1) an endless sequence of theories or (2) different parts of physics describing difference aspects of reality, with no overall ability to provide comprehensive explanations. If the first alternative (1) is correct, and I believe that it is, it is no deterrent to the endless quest of metaphysics and science to understand our world. We have now circled back to where we began. Just as Newton, or Aristotle before him, provided a coherent paradigm that gave unity and direction to physics for a time, so will the unification of relativity and quantum theory if and when it occurs. A 'final' theory simply means the final step of unification in the current attempt to develop a comprehensive paradigm, not a final theory of physics. Any physical theory will face the inevitable decay of the framework and the shock of a scientific revolution. The second alternative (2) presents a more challenging claim, but if we can make the very plausible assumption that nature itself is unified, history would appear to be on the side of those who have achieved unified theories, themselves parts of other more general theories.

Notes

CHAPTER 1

1. For the majority of contemporary textbook, dictionary and encyclopedia sources on metaphysics, entries for events uniformly treat them as dependent on substances. Even when Quine's views are discussed, it is not at all clear that there is another, competing conception in which events function as the basic particulars. One exception is Campbell (1976: 32–5).

2. See for example Whitehead (1919, 1920, 1922), Russell ([1927] 1934) and Broad (1923). Whitehead and Russell both delivered their theory of events as the Tarner Lectures at Trinity College in the University of Cambridge. Whitehead identifies the purpose of the Tarner lectures as 'the Philosophy of the Sciences and the Relations or Want of Relations between the different Departments of Knowledge' (1920: v). His 1919 lectures were published as *The Concept of Nature* in 1920. Russell's 1926 lectures were published as *The Analysis of Matter* in 1927. Broad says in the preface of his *Scientific Thought* that Whitehead and Russell led the way by combining constructive fertility, penetrating critical acumen and technical mathematical skill, while he claims 'the humbler (yet useful) power of stating difficult things clearly and not too superficially' (1923: 6). Whitehead, Russell and Broad do not use the term 'event ontology' to characterise the position they are defending, but it is nonetheless clear that they are all supporting the thesis that events are basic and best serve as an ontological foundation for physics. Whitehead uses the terms 'pan-physics', 'philosophy of natural knowledge' and 'philosophy of organism' to describe the views he developed in his middle and late periods.

3. See for example Stapp (1979, 1982, 2011) and Epperson (2004). Also see Papatheodorou and Hiley (1997), who argue that a successful theory of quantum gravity requires a radical change in our conception of physical reality. In Chapter 6, I take up these views.

4. As analytical philosophy has become preoccupied with its own history, there has been a certain amount of dispute over what constitutes the origin of the analytic tradition. Dummett (1994), for example, has claimed that

post-Fregian analytic philosophy is distinguished by the foundational role of the philosophy of language, particularly by the search for a viable theory of meaning. This has the odd consequence of denying Russell membership to the tradition. Hacker (1996) sees a schism over the relationship between philosophy and science. Those who see a sharp distinction between the two regard philosophy as a quest for human understanding via conceptual clarity. Wittgenstein, Moore, Ryle, Austin and Strawson, then, are the quintessential analytical philosophers because they eschew the orientation of the scientific investigator as a guideline for philosophical work. According to this view, those who see philosophy and science as engaged in a collaborative effort for truth about the world have abandoned the true analytic ambition, for example Russell and Quine. Also see Strawson (1990).

5. Even these views seem relatively tame compared with cutting-edge string theory employing as many as ten dimensions to explain all fundamental particles and forces, or bane theory which explains how strings attach to membrane-like surfaces in higher dimensions.

6. For an excellent exchange on this subject see Susan Haack (1998) and Strawson's reply, which confirms his commitment to the idea that philosophy should be concerned, among other things, with the structure of our common thinking and that he is 'unconvinced of the combined conceptual-empirical character of philosophical enquiry' (1998: 65).

7. Anthony Quinton, for example, has remarked: 'Outside the sequestered province of the cult, Whitehead is regarded with a measure of baffled reverence, mingled with suspicion.' While Whitehead was clearly made of 'the right stuff', says Quinton, his philosophical writings have been utterly incomprehensible to the general philosophical community (1985). One reason for this is the interdisciplinary character of Whitehead's work. He was not trained as a professional philosopher, but rather came to philosophy late in life after having made significant contributions to mathematics, logic and physics. In other words, Whitehead was not constrained by the specialised training that divides disciplines, but his intellectual milieu at Cambridge was always deeply philosophical (see Lowe 1985, 1990). More recently, Peter Simons' comment that Whitehead 'obtrudes anomalously in the history of twentieth century philosophy' resonates with Quinton's view. But unlike Quinton, Simons further notes that Whitehead 'is in the same grand scientific metaphysical tradition as Russell and Quine, but he has a mind of his own, and if not in the content of his work, in his basic philosophical attitudes I think he is dead right' (2003: 663, 666). In this book, I hope to make Whitehead's process metaphysics intelligible to a wider audience of philosophers and scientists by situating his thought in that scientific metaphysical tradition and by drawing comparisons with Russell, Broad and Quine.

8. While Quine wrote his doctoral dissertation, 'The logic of sequences: a generalization of *Principia Mathematica*', at Harvard University under

Whitehead's direction in 1931, he does not acknowledge any philosophical influence from Whitehead. Rather, he cites Carnap and Russell as the main inspirations of his early development. When considering a topic for a term paper in Whitehead's seminars, he said he 'took refuge in his relatively mathematical material on "extensive abstraction"' and while he responded little to Whitehead's philosophy courses, 'retained a vivid sense of being in the presence of the great' (1986: 9–10). When Quine overlapped with Whitehead at Harvard, Whitehead was deeply involved in working out the details of his metaphysics of process – a philosophic vision that ran against the positivistic spirit of the times. Just prior to Whitehead's metaphysical period, however, he devoted his attention to a general ontology of physics and a theory of natural knowledge. This is where Whitehead and Quine connect. Quine's dissertation was subsequently published under the title, *A System of Logistic* (1934). For his affinities to Whitehead, see McHenry (1995a, 1997). For Quine's reply, see Quine 1997.

9. 'Naturalised metaphysics' as I have used the term is closely associated with the Quine-Dewey project of naturalising knowledge, meaning and mind. See McHenry (1992, 1995a, 1998).

10. On this point there is a difference between Whitehead and Quine since Quine is well known for his defence of scientism, namely the view that physical science is our only basic source of knowledge. Whitehead would agree that that our ontology must take account of advancing physics but not that physical science is the only basic source of knowledge. He did not, for example, explain consciousness by psychological behaviourism. Susan Haack, in this connection, says that Quine's 'scientific commitment' exposes him to the objection that 'his scientistic leanings transmute the continuity theme into a kind of scientific imperialism in which, at best, philosophical inquiry is turned over to the sciences, or, at worst, philosophical inquiry is abandoned altogether in favor of scientific inquiry' (1998: 52).

11. For this formulation of the problem see Nicholas Maxwell (2006) who attributes the idea to Chris Isham (1993, 1997). The notion that there are two different conceptions of time in quantum theory and general relativity is also recognised by Henry Stapp as a fundamental problem of the final unification (1982).

12. Steven Weinberg, for example, says: 'Physicists do of course carry around with them a working philosophy. For most of us, it is a rough-and-ready realism, a belief in the objective reality of the ingredients of our scientific theories. But this has been learned through the experience of scientific research and rarely from the teachings of philosophers.' Even the philosophy of science, he claims, fails 'to provide today's scientists with any useful guidance about how to go about their work or about what they are likely to find' (1994: 167). Stephen Hawking also writes: 'The people who actually make the advances in theoretical physics don't think in the categories

that the philosophers and historians of science subsequently invent for them' (1993: 43).

13. The metaphysical and methodological aspects of unification are a central focus of Nicholas Maxwell's aim-oriented empiricism. See Maxwell 1998.

CHAPTER 2

1. Aristotle's views on substance are mainly found in *Metaphysics*, 1028 and 1029, and *Categories*, 3a–4b (1941: 783–7, 10–14).

2. Aristotle's work brings to full completion a line of thought that began with the earliest philosophers who regarded the first principle as corporeal, as, for example, he states in Book I of his *Metaphysics*, 987a (1941: 700).

3. Another good example of an analytic approach to events in the tradition of descriptive metaphysics is Jonathan Bennett's *Events and their Names*. While Bennett advances a doctrine of supervenience according to which events are dependent on substances and properties, and not vice versa, he also explicitly claims that he has no opinion to venture about the reversal of the supervenience. He says: 'events are supervenient on substances and properties, unless the supervenience runs the other way because tropes are more fundamental than substances and properties'. Events, he says, are tropes; the only difference is that the latter are slightly more general than the former (1988: 16). Bennett does not attempt to refute the view that makes events the primary ontological units, nor indeed the only concrete particulars, even though later in his book he considers Quine's view of physical objects as four-dimensional and remarks that Quine is untroubled 'by the intuitive implausibility of his view of events or to go through the moves I have supposed "the Quinean" to make so as to reconcile his metaphysic with what intelligent people in the street think and say' (ibid.: 113). As he makes it clear at several points, Bennett's explicit intention is to analyse our event concept and determine how it does its job in ordinary language (ibid.: 20, 118). So, in accordance with the approach taken by Aristotle, the status of events is bound by the straightjacket of ordinary language and the proverbial man on the street.

4. For example, Horgan (1978). Also see Thalberg (1985) for an analysis and evaluation of the position offered by the anti-event metaphysicians or what he calls 'no-event metaphysicians'.

5. The second question regarding identity and individuation has largely focused on the search for an adequate criterion of identity. Assuming that there are events, what are the necessary and sufficient conditions for deciding whether two events *e* and *e'* are identical or not? If Brutus stabs and thereby kills Caesar, is his stabbing him the same event as his killing him? An answer to this question is thought to be important because it would greatly clarify the nature of events and their relation to other entities such as properties, relations and classes. Others claim that no such strict criterion of identity is possible: events are merely individuated conveniently

by our descriptions of them. See especially the exchange between Quine and Davidson: Quine (1985a: 162–71), Davidson (1985: 172–6). Also see Bennett (1988: 122–8) and Unwin (1996).

6. In this regard, descriptive metaphysicians are guilty of a type of hubris, an anthropocentricism perhaps best captured in Hume's quip: 'But the life of a man is of no greater importance to the universe than that of an oyster' ([1741, 1742] 1963: 590). Hume's point, slightly modified for the present purpose, can be restated: the structure of human speech is of no greater importance to the universe than the murmur of an oyster.

7. Descriptive metaphysicians might reject this particular characterisation of substance as an unchanging substratum but the force of Whitehead's criticism against the linguistic origin of the substance concept nonetheless still remains intact. For a more detailed exposition of Whitehead's argument, see McHenry (1990).

8. The bifurcation of nature has a long history beginning with Democritus' early distinction between atoms and the void and our illusory sensations resulting from the motions of the atoms, and continuing with Galileo, Descartes, and Locke's versions of primary and secondary qualities, and Kant's distinction between the noumenon and the phenomenon.

9. Scholastics, for example, tried to make sense of perception using Aristotle's theory of hylomorphism. When we see an object, our 'sensible faculties' are imprinted with the actual form of the object and thus the form is known in abstraction from the matter. This 'direct realism' or 'naïve realism' was unintelligible to the moderns who then introduced 'indirect realism' or the representational theory of perception that distinguishes between objects as they really are and objects as they appear to us. But, for Whitehead, this only produced an epistemological morass because it entails that the objects of our knowledge exist beyond the veil of phenomena and therefore remain forever unknowable.

10. Whitehead, in a related objection to substance ontology, argues that the human intellect tends to ignore the fluency of process by 'spatializing the universe' ([1929] 1978: 209). Substance, as a metaphysical substratum or the unchanging subject of change that was emphasised by Aristotle, falls under this objection as does the idea that in human perception we naturally focus on the clear and distinct sense from visual perception and neglect the more rudimentary perceptions of our bodily experiences, namely the process of inheritance (ibid.: 117). In this manner, the subject-predicate distinction of our ordinary conceptual scheme reinforced spatialisation and gave metaphysical priority to an accident of evolution – that we happened to evolve with these acute modes of perception. In Whitehead's explanation of the process of becoming he postulates that the vast majority of actualities that comprise nature experience the emergence from the immediate past, but only the select few in higher organisms experience the sort of perceptions that arise in visual experience (ibid.: 212). In other words, the capability of the intellect to spatialise and to arrest

for contemplation and analysis is rare in nature; what is dominant in the experience of human beings should not be confused with the more fundamental processes in the greater number of organisms throughout nature. See Chapter 7 for further treatment of this topic.

11. This is the focus of Roger Gibson's excellent exposition of Quine's philosophy. See especially (1982: xvii–xviii, 31–62).

12. See C. I. Lewis, 'A pragmatic conception of the a priori' (1923: 169–77). Quine explicitly acknowledged Lewis's influence on him in Barrett and Gibson (1990: 292).

13. This signals Quine's disassociation from what he calls 'card-carrying pragmatists' (1981a). While his naturalism is consistent with the classical pragmatists' rejection of certainty, he is not espousing pragmatism in the fashion of James and Dewey, for whom the test of ideas generally is their cash value in experience, including aesthetic, moral and religious experience. For Quine, this attitude is one that is too hospitable towards comfortable beliefs such as wishful thinking and the like (ibid.: 32–3). Pragmatism has a limited function in Quine's system, that is within the context of theory selection in science. For further clarification, see McHenry (1995a).

14. Also note here how Mei's charge against Strawson echoes Whitehead's complaint that it was the imperialistic and ethnocentric Greeks who unjustifiably took the 'historical accident' of their grammatical structure for a universal conceptual scheme.

15. This was a focus of Davidson's famous paper, 'On the very idea of a conceptual scheme', in which he denies that we can make sense of alternative conceptual schemes, that is ones in which the language of that scheme cannot be translated into our language. So, the question of whether alternative conceptual schemes exit is transformed into the question of whether non-intertranslatable languages exist. See Davidson (1984). As we saw in Chapter 1, Davidson argues that science is a 'suburb' of our commonsense conceptual scheme; it can add to it but it cannot subtract. This is especially true when it comes to dethroning the central place of substance in that scheme. Given the radical nature of the paradigm change required by quantum mechanics, electromagnetic theory and relativity theory, the idea that science is a suburb of our ordinary thought no longer seems plausible. In the following chapters, part of my aim is to demonstrate exactly how radical this change of paradigm has been.

CHAPTER 3

1. For further discussions on Maxwell's influence on Whitehead, see Lowe (1966), Fagg (1997), and Athearn (1997).

2. Bertrand Russell reports that Maxwell's *A Treatise on Electricity and Magnetism* was the subject of Whitehead's Fellowship dissertation at Trinity College, Cambridge, in 1884 ([1959] 1995: 33). Victor Lowe

discusses this choice of topic in connection with Whitehead's mathematics teacher at Cambridge, W. D. Niven, who was a student of Maxwell and edited Maxwell's *Scientific Papers*. No copy of Whitehead's dissertation survives (Lowe 1985: 94–6, 106–7).

3. For an exposition of Maxwell's equations, see Feynman, Leighton and Sands (1964: chapter 18). For a more general exposition of the development of Maxwell's thought, see Harman (2001).

4. Nicholas Maxwell (1993) provides an illuminating exposition of Einstein's methodology of resolving conflicts between well-established fundamental theories and formulating a new unifying theory in the creation of the special and general theories of relativity. In the first instance, this involved the unification of Newtonian mechanics and Maxwellian electrodynamics in the special theory and, then, in the second instance, the unification of Newton's theory of gravitation and special relativity in the general theory. Finally, Einstein sought unsuccessfully to unify general relativity and quantum theory.

5. While there is some dispute among historians of science as to whether Einstein was directly influenced by the results of the Michelson-Morley experiment, his 1905 paper on special relativity, 'Zur Elektrodynamik bewegter Körper', translated into English as 'On the electrodynamics of moving bodies', does refer to unsuccessful attempts to measure the motion of the earth relative to the ether, see Einstein (1923: 35). The Michelson-Morley experiment is also discussed in Einstein ([1920] 1962: 53, 147).

6. More recently, Tim Maudlin writes: 'Once we accept events as the basic elements of our spatio-temporal ontology, our job is to specify the physical structure of these events: the geometry of space-time' (2012: 60–1). In addition, see John Norton who also affirms events as the basic building blocks of the space-time structure (2014). Theodore Sider, on the other hand, who defends four-dimensionalism, does not explicitly defend an event ontology in spite of his concerns about temporal parts of objects (2001: xiii, 1–10).

7. Whitehead criticised Einstein's interpretation of the metric of space-time in terms of rigid rods, periodic clocks and the constant transmission of light. For Whitehead, rigidity, periodicity and constancy presuppose the uniformity of space-time that is rooted in congruence. The possibility of measurement depends on exact congruence between regions of space. Otherwise, we should not have any standard for determining what we mean by certain distances. With A. S. Eddington's confirmation of Einstein's general theory of relativity in 1919, however, it is clear that Einstein's heterogeneous space-time structure won the day and Whitehead's critique went largely unnoticed. Whitehead's logical point was overshadowed by the strength of the empirical evidence that was the foundation of Einstein's theory. See McHenry in Lowe (1990: 126–7).

8. Also see Whitehead's chapter VII on the quantum theory in *Science and the Modern World* (1925).

9. On this point, see especially Folse (1974: 43–6).
10. At the time of writing, the elementary particles of the standard model include: six 'flavours' of quarks: up, down, bottom, top, strange and charm; six types of leptons: electron, electron neutrino, muon, muon neutrino, tau, tau neutrino; twelve gauge bosons: the photon of electromagnetism, the three W and Z bosons of the weak force, and the eight gluons of the strong force; and the Higgs boson. Stephen Hawking argues that the search for elementary particles cannot be endless since gravity provides a limit. He writes: 'If one had a particle with an energy above what is called Planck energy, ten million million million GeV [giga-electron-volt] . . . its mass would be so concentrated that it would cut itself off from the rest of the universe and form a little black hole' (1988: 167).
11. Quine in 'Whither physical objects?' comes to the same view when he takes up the problem of identify and individuation of physical objects. As physics continues to probe the subatomic realm, the problem of finding clear criteria for identify increasingly applies to particles. See (1976).
12. Salam, Glashow and Weinberg, for example, unified the weak nuclear force with the electromagnetic force in their theory of the electroweak force just as Maxwell had unified electricity and magnetism. 'Although these two forces appear very different at everyday low energies, the theory demonstrates that they are two different aspects of the same force. Above the unification energy, on the order of 100 GeV, they would merge into a single force. Thus if the universe is hot enough (approximately 10^{15} K, a temperature exceeded shortly after the Big Bang) then the electromagnetic force and weak force will merge into a combined electroweak force' (Wikipedia entry for 'electroweak interaction', accessed December 2014).

CHAPTER 4

1. Dorothy Emmet claims in her Cambridge Philosophers Lecture, 'Whitehead', that one of the themes that runs throughout Whitehead's thought is his interest in the relation between logico-mathematical schemes and the rich, complex world of experience (1996: 103, 112). Victor Lowe calls this Whitehead's doctrine of 'the rough world and the smooth world' (1966: 180–1). Aside from the method of extensive abstraction where classes or sets aid in the development of abstractions, another good example of how the advances in logico-mathematical schemes have provided tools for the organisation of thought in cosmology and metaphysics was the development of the logic of relations initially pioneered by Charles Peirce and Augustus DeMorgan and advanced by Whitehead and Russell in *Principia Mathematica* (1910–1913). Paraphrasing Russell, the logic of relations gave thought wings whereas the categorical logic of Aristotle put it in fetters ([1914] 1956: 53). The immediate consequence was crucial for mathematics and

metaphysics since it allowed for all sorts of distinctions of order and sense for quantitative differences, for example transitive and asymmetrical relations.

2. For Whitehead's theory of extensive abstraction, see (1919: 101–64; 1920: 74–98). Also see Lowe (1966: 62–71) and McHenry (1990: 119–23). The basic idea is that the elements needed to construct the geometry of space-time can be derived by a process of abstraction from classes. The 'abstractive element' – point, line, plane and so forth – is defined as the class of all equivalent abstractive sets of the same type. The abstractive element replaces the point or instant as an entity radically different from anything known in experience; it is rather understood as an ideal limit of diminution of extensions. For example, no one has an experience of a point or an instant, but we can experience a progressive refinement of perceptible volumes or durations to ideal limits thus defining points and instants. Geometrical entities or instants of time are then viewed as logical functions of extension instead of actual particulars on a par with events or objects, yet they do the same mathematical work that is required for physics. For example, 'force at a point' or 'configuration at an instant' is now understood in terms of ideal entities derived from converging series of extensive regions.

3. There are actually eight Categories of Existence in Whitehead's Categoreal Scheme: Actual Entities (or Actual Occasions), Prehensions, Nexūs, Subjective Forms, Eternal Objects, Propositions, Multiplicities and Contrasts. Of the eight categories, he says, actual entities and eternal objects stand out with a certain extreme finality whereas the others have a certain intermediate character ([1929] 1978: 22). For our purpose we need not include all of the complexity of the categories of existence. The dualism between events and properties (or actual occasions and eternal objects) remains basic.

4. Throughout *Process and Reality* there is little suggestion of Whitehead's method of generalising from our psycho-physiological embodied experience. But compare, for example, an earlier description of this procedure with a later one. In *Science and the Modern World* he says:

> In this sketch of an analysis more concrete than that of the scientific scheme of thought, I have started from our own psychological field, as it stands for our cognition. I take it for what it claims to be: the self-knowledge of our bodily event. I mean the total event, and not the inspection of the details of the body. This self-knowledge discloses a prehensive unification of modal presences of entities beyond itself. I generalise by the use of the principle that this total bodily event is on the same level as all other events, except for an unusual complexity and stability of inherent pattern . . . if you start from the immediate facts of our psychological experience, as surely as empiricist should begin, you are at once led to the organic conception of nature . . . (1925: 91–2)

Much the same idea is expressed in *Adventures of Ideas*, where he says:

> if we hold, as for example in *Process and Reality*, that all final individual
> actualities have the metaphysical character of occasions of experience, then on
> that hypothesis the direct evidence as to the connectedness of one's immediate
> present occasion of experience with one's immediately past occasions, can be
> validly used to suggest categories applying to the connectedness of all occasions
> in nature. (1933: 284)

Both statements make it clear that Whitehead's concepts of the actual
occasion and prehension originate in introspection on concrete experience.
He was particularly careful not to confuse the 'clear and distinct' elements
in our conscious experience with what is basic. This is where he makes a
radical departure from traditional empiricism. It is rather the vague and
inarticulate feelings of our total bodily experience that he believes capture
the essence of becoming. This might seem like Whitehead is anthropo-
morphising nature by generalising from human experience, but he makes
it clear that one could arrive at the organic conception of nature and the
event ontology from either psychology and physiology equally, or from
modern physics; in his case, he says it was mathematical physics that was
the primary factor by which he came to his convictions (1925: 189–90).
I explain Whitehead's view of experience with regard to perception and
causation in Chapter 7.

5. Russell's main maxim – to substitute constructions out of known entities for inferences to unknown entities – in his philosophy of science and
   epistemology was greatly influenced by Whitehead's novel apparatus of
   the method of extensive abstraction. In fact, he credits Whitehead for his
   awakening from 'dogmatic slumbers'. Russell saw the method of extensive
   abstraction as an application of Occam's Razor in physics, for one need
   not assume the abstract constructions as part of the furniture of the world
   ([1959] 1995: 77–8, 81). Points, lines and planes, for example, are not
   included among the entities recognised in our ontology but rather are
   viewed as abstractions or ideals of pure thought that are derived from
   experience.

6. Commentators on Russell's philosophy seldom discuss his event ontology.
   Even Quine in 'Russell's ontological development' fails to mention the
   place of events in Russell's ontology. What does get attention is how logic
   shaped Russell's ontology until finally we arrive at the construction of the
   external world from sense data (1981b: 73–85). The exceptions include
   Nagel (1946) and Fritz (1952).

7. Russell reaffirmed his position on events in 1946 when he replied to
   Nagel in the volume devoted to his thought in *The Library of Living
   Philosophers* series (1946: 701, 705).

8. In his foreword to Quine's *A System of Logistic*, Whitehead seems to have
   anticipated the general point of Quine's theory when he wrote that: 'logic
   prescribes the shape of metaphysical thought' (1934: x). For example, in
   Principia Mathematica, Whitehead and Russell put Φ and Ψ in quantifiers
   – for example, '$(\exists\Phi)(x)\Phi x$' – such that attributes become variables in

their own right. As Quine says, 'The effect of letting "Φ", "Ψ", etc. occur in quantifiers, now, is that these letters cease to be fragments merely of dummy matrices "Φx", "Ψy", etc., and come to share the genuinely referential power of "x", "y", etc.' (1941: 145). Throughout Whitehead's philosophy, he proposes an ontology that is dualistic in the sense of recognising both individuals and properties. In Quine's logic, however, individuals and classes are properly quantified over – for example, '(∃x) (Φx)' – but are not attributes (1953: 102–3; 1969: 91–2). This has an important consequence for 'the shape' of Quine's ontology in that he does not recognise properties or attributes in addition to objects or individuals. As he puts the point: 'Properties were vaguely assumed in *Principia* as further denizens of the universe, but they serve no good purpose that is not better served by classes, and moreover they lack a clear criterion of identity' (1985b: 85).

9. In this essay, 'Whither physical objects?', Quine considers the evidence from recent developments in physics for his view of physical objects. From the philosophical perspective, Quine's thesis of the inscrutability of reference confirms these findings. It is indeterminate what objects the singular terms, pronouns and bound variables of our true sentences refer to.

10. This question was raised in McHenry (1997: 9–10). Quine responded to McHenry, with the purported aim of clarifying both his affinities, as well as his disagreements, with Whitehead's dualistic ontology of individuals and properties (1997: 13) The paper was subsequently reprinted in two books: *Process and Analysis: Whitehead, Hartshorne and the Analytic Tradition*, edited by George Shields (2003), and *Quine in Dialogue*, edited by Dagfinn Føllesdal and Douglas Quine, (Quine 2008a). In his 'Response to Leemon McHenry,' Quine mentioned and partially discussed three criticisms of properties: (1) properties lack a 'clean-cut principle of individuation'; (2) there is nothing that can be 'explained in terms of properties that could not be explained equally well in terms of classes'; and (3) spatio-temporal coincidence is a wholly satisfactory principle of individuation, so properties are not needed (1997: 13–14). For a detailed reply to these criticisms, see Holmgren and McHenry (2012).

11. Quine's brand of extensionalism rejects properties (or attributes) in favour of classes, but includes abstract entities, provided that they have a clear principle of individuation. Nominalism, on the other hand, is typically understood as the rejection of universals in favour of an ontology of exclusive individuals. Everything real is particular and abstract universals are nothing but names; the mistake of dualists, according to nominalists, is their reification of general concepts. As Quine puts it: 'Discourse adequate to the whole of science can be so framed that nothing but particulars need be admitted as value of variables' (2008b: 16). But Quine himself was not a nominalist in this sense. It is important to be clear about two kinds of nominalism in metaphysics: (1) the view that there are no abstract objects, and (2) the view that there are no universals when universals are construed

to be properties. Quine is a nominalist in the latter sense but not the former. As he was motivated largely by the nominalist's principle of parsimony, or as he put it in his classic 'On what there is', 'a taste for desert landscapes', he sought to keep his ontology to a bare minimum (1953: 4). Nominalism, for him, is an ideal for ontological construction, one that seeks a theory about what there is, no more, no less. In this manner, the metaphysician attempts to avoid empty theorising and perplexing questions about the relation between universals and particulars. But as much as Quine aspired to nominalism in the sense of (1) above, he was forced to admit classes to his ontology, and to reluctantly concede to a form of Platonism, his main reason being that science would be impossible without recognising their service to the formulation of quantitative laws, numbers being reducible to sets (1981b: 13–14).

12. Quine is not included in this response to the criticism that all observation is theory laden and that any attempt to start afresh from sense perception is itself loaded with conceptual baggage since he clearly recognises and accounts for this objection in his holism and in his theory of underdetermination (see Chapters 1 and 2).

CHAPTER 5

1. Whitehead's theory of cosmic epochs is not to be confused with the *a priori* investigations into the nature of alethic modalities in the fashion of Gottfried Leibniz, Saul Kripke and Alvin Plantinga. Nor is he advancing an argument for modal realism, in the manner of David Lewis, that all possible worlds are real. Whitehead's cosmic epochs are more in accordance with what cosmologists refer to as parallel worlds or a multiverse comprised of a plurality of universes. The cosmologists, in contrast to the philosophers, are engaged in *a posteriori* investigations into the nature of the physical universe and scientific theorising arising from these investigations. Possibilities in Whitehead's metaphysics are treated in his theory of eternal objects.

2. There is a difficulty here of understanding in what sense the different cosmic epochs stand in relation to one another without some single measurement of time and space – or in a Superspace – for all *cosmoi*. The view seems to assume a sort of Newtonian absolute time and space, but such conceptions can only make sense relative to some individual epoch. This problem is also recognised by multiverse theorists. See for example Rees (2001: 170).

3. In this scenario, once God created the physical world, the structure (including the laws of nature) was set. Laws of nature are necessary in the sense that they were created by God who is perfect, good, and therefore not deceiving; once set they do not change with the passage of time. The laws are universal in the sense that there is only one extended substance and the laws apply throughout this one universe. See Descartes ([1644]

1911: Part 2, XXI, XXII, XXIII, XXXVI, XXXVII, LXIV, Part 3, I) and Kneale (1949). Also see Popper (1959: 430–1).

4. When Whitehead's *Process and Reality* was first published in 1929, the only known particles were electrons and protons, the only known forces gravity and electromagnetism.

5. The one exception in the physics literature is Barrow and Tipler's *The Anthropic Cosmological Principle* (1986: 192–3).

6. The term 'multiverse' was first used by William James in 1895. See James ([1897] 1956: 43). In cosmology, however, the idea has been advanced at least as far back as Nicholas of Cusa in the fifteenth century and was espoused by Giordano Bruno in his *On the Infinite Universe and Worlds* (1584). Bruno's proposal was just one of the heresies that brought him before the Inquisition and had him burned at the stake at the Campo dei Fiori in Rome in 1600.

7. When most commentators on Whitehead's metaphysics discuss this topic they seem to assume there is only one cosmic epoch at a time, that is, that the *cosmoi* form a single-line sequence. One exception is Rem B. Edwards who, in his 2000 article in *Process Studies*, notes that Whitehead affirms contemporary cosmic epochs spread out in an infinite Superspace (2000: 87).

8. Aleksander Wolszczan and Dale Frail carried out astronomical observations from the Arecibo Observatory in Puerto Rico which led them to the discovery of two planets orbiting the pulsar PSR B1257+12 in 1992. This was the first confirmed discovery of planets outside the solar system. See Wolszczan and Frail (1992).

9. One such promising observational test for the eternal inflation hypothesis has been conducted using cosmic microwave background data. Bubble collisions with other universes produce inhomogeneities in our inner bubble, thereby resulting in observable signatures that are detected in the cosmic microwave background. See Feeney et al. (2010).

10. See Weinberg (1994: chapters IX and X) and Greene (2003: chapters 14 and 15). Beyond the unification of the standard model and general relativity, if a final theory is understood in the context of superstring theory or M-theory (encompassing the multiverse hypothesis), there is then even more basis for the comparison with Whitehead's theory of cosmic epochs, for the ultimate theory would extend beyond the unification of forces in our limited cosmic epoch. See especially Greene (2003: 368–9).

CHAPTER 6

1. Stapp argues that Bohr's view of the quantum theoretical formalism is largely in agreement with William James' pragmatism. The most important connection between the Copenhagen interpretation and Jamesian pragmatism is the epistemological imperative to explain merely the possible relations between the multifold aspects of experience and reject

any attempt to discover the underlying essence of phenomena. See Stapp (1972). The Copenhagen interpretation of Bohr and Heisenberg need not be interpreted as anti-realist in that they supposedly denied the existence of an objective external reality but rather that they simply regarded the formalism of quantum theory as a heuristic device. In this sense the disagreement with Einstein was not about the existence of external reality but rather about the manner in which the external reality was meant to be described. See Krips (1987: 1, 23–4).

2. I owe this distinction to Malin (2001: 39).

3. For Maxwell's own solution to what he calls 'the fundamental problem concerning the nature of the quantum world', namely, the wave/particle problem, see Maxwell (1988, 1994, 1998). He says for example: 'The world is made up of propensitons, fundamentally probabilistic objects which, in appropriate conditions, evolve into superpositions of states and, as a result, smear out spatially in a way that is unlike anything encountered in deterministic classical physics' (1994: 353–4).

4. In this connection, Whitehead, in a purple passage, notes: 'But in the present-day reconstruction of physics fragments of the Newtonian concepts are stubbornly retained. The result is to reduce modern physics to a sort of mystic chant over an unintelligible universe' ([1938] 1968: 136). From the context, it is clear that he has in mind the replacement of empty space with the field of force, one of incessant activity that he argues eliminates matter from the picture.

5. Insofar as quantum field theory, quantum electrodynamics, quantum electroweak theory, quantum chromodynamics and the standard model all employ the Copenhagen interpretation of quantum theory, the same criticism holds. All are interpreted as forms of instrumentalism.

6. Aside from Stapp, the quantum theorist, David Bohm, has proposed an ontology for quantum theory that is partly inspired by Whitehead's event theory (Bohm 1980: 48, 1986: 183, 186). Whitehead's extensive continuum of events is understood to be the ontological basis for what Bohm calls 'implicate' order or the 'enfolded' order that functions as the fundamental order of reality (1980: xv). This, he argues, is the underlying structure of quantum phenomena which the orthodox theory fails to explain. The implicate order of nature involves pre-space structures created at the micro-level which give rise to the 'explicate' or 'unfolded' order – time, space and matter – at the macro-level and by which we observe and measure physical phenomena such as periodic vibrations of subatomic particles, oscillations of electromagnetic fields, entropy and ultimately cosmological processes. The mathematics of quantum theory deals with the implicate pre-space. Bohm takes Whitehead's ontology of actual occasions or 'moments' as he calls them as the fundamental basis to construct his theory but his view of the implicate order is more akin to the view Whitehead developed in his pan-physics rather than his metaphysics of process (Bohm 1980: 207).

A similar project has been undertaken by Frank Hättich who has provided a sustained analysis of quantum field theory in terms of Whitehead's event ontology by demonstrating how Whitehead's metaphysical scheme can be matched to the mathematics of algebraic quantum field theory. Hättich has also suggested how Whitehead's ideas need to be revised in order to bring them up to date with contemporary theoretical physics (2004). Even more recently, Michael Epperson and Elias Zaffiris have embraced an event-based interpretation of quantum mechanics that is compatible with Whitehead's mereotopological theory. They argue that the 'primacy of objective quantum events is . . . the critical starting point for any attempt to provide a viable ontological interpretation of the standard formal framework of quantum theory' (2013: xvi). Also see Shimony (1965), Stapp (1979, 2011), Papatheodorou and Hiley (1997), Malin (2001), Eastman and Keeton (2004) and Epperson (2004).

7. Interestingly enough Heisenberg explicitly acknowledged that the modern interpretation of atomic events has turned away from the materialistic philosophy of classical physics, and has come 'extremely near to the doctrines of Heraclitus' (1958: 63). What is less clear, however, is whether he would have espoused any sort of micro-realism of the sort proposed by Whitehead. The passage quoted above reaffirms that the probability function must be interpreted epistemologically.

8. Compare, for example, Geoffrey Chew's discussion of the multiple levels of fundamental process, including 'pre-events'. Chew has attempted to explain how material reality emerges from pre-events using Planck time scales within historical reality. Working with a Whiteheadian concept of process, he construes matter to be 'regular localized repeating patterns of large numbers of occasions. The model locates an impulse at each pre-event; spacing between successive pre-events is on a scale of Planck time – roughly 10^{-43} seconds.' In his view, the individually localised pre-events form the base of the system upon which matter is structured beginning with the minimum duration of an elementary particle 10^{-25} seconds, then the period of an atom 10^{-15} seconds, and so on (2004: 87).

9. See especially Čapek's 'Time-space rather than space-time' (in Čapek 1991: 324) and 'The dynamic structure of time-space' (in Čapek 1961: 214). Tim Maudlin offers an alternative conceptual apparatus in the theory of linear structures in response to the complaint that the traditional geometry of relativity theory 'spatializes time'. There is a choice involved as to which mathematical tools one uses to represent the geometry of space-time and a new foundation can be built on the notion of a line or directed line that 'temporalizes space'. See Maudlin (2010). Also see Cahill, who defends a process physics against the four-dimensionalism of general relativity (2005).

10. Stapp provides useful diagrams of the relativistic and non-relativistic quantum field theories in his figures 13.1 and 13.2. In the former, a curvy line on the bottom of the diagram represents the space-like

three-dimensional surface 'now' that separates the space-time region of the fixed past from the potential future. The events in a process of becoming are different space-time regions on this curvy line, each with its own perspective on the past and each pushing the boundary surface 'now'. See (2011: 92, 93).

11. Compare, for example, Papatheodorou and Hiley (1997: 266) and Stapp (1982: 382).

12. Some eternalists have explained the passage and direction of time in a quasi-Kantian manner by distinguishing between a subjective experience of time's flow and the objective reality of an unchanging four-dimensional reality, that is, a *manifest time* and a *scientific time*. Adrian Bardon, for example, argues: 'As it is with what we call *now*, the *direction* of time means something to us, but it appears to mean nothing to the universe' (2013: 122). Gödel also espoused this position when he interpreted the results of relativity theory. See Yourgrau (1991: 38). This strategy accepts a variation on the bifurcation of nature that we rejected in Chapter 2 mainly because it turns our experience into a grand illusion. In fact, it is unclear why there should be any direction to time at all if we are to take seriously the full implications of the block universe, namely full spatial symmetry in four dimensions.

13. Christian (1959), Sherburne ([1966] 1981), Johnson (1976) and McHenry (1992, 2000) have all defended the traditional interpretation where Whitehead appears to advance some form of presentism.

14. The possibility of time travel is most often discussed in connection with the Einstein-Minkowski four-dimensional universe. Within this view, since each event in the space-time block is eternally present, the idea of travelling in a time machine to 1955 to visit Einstein made science fiction closer to a genuine physical possibility, time being of course more akin to space. Kurt Gödel, for example, entertained this possibility, interestingly enough while he was a colleague of Einstein at the Institute for Advanced Studies in Princeton. The general theory of relativity allows for an R-universe where the compass of inertia rotates everywhere relative to matter and there are no co-moving frames that form a single, consistent, cosmic time. In such a world there are closed, future-directed, time-like curves (CTCs) where we can 'travel' in time by heading into our causal future and eventually arrive at our causal past or present. See Yourgrau (1991: 20, 46). However, with the growing block universe, it is unclear what one would encounter if one were to travel to the past. The past event is really there in the block but no longer present in itself.

15. Whitehead considered God, an actual entity in process with the world, to be an essential component of his metaphysics, especially in connection with his explanation of the past. This notion is subject to the objection that God is an *ad hoc* modification introduced to save the present from passing into complete oblivion. See McHenry (1992) for further treatment of this topic in connection with what Whitehead calls 'the consequent

nature of God'. As an aside, in opposition to the classical theology in which God is conceived as a divine substance, in Whitehead's neo-classical theology God is an everlasting event.

16. On this score, one can see why eternalists such as Spinoza and Sprigge accept a metaphysical monism that identifies God and nature. For these philosophers, an event never ceases to be present in itself (or subjectively immediate in Whitehead's terminology). For presentists and proponents of the growing block universe, this is too high a metaphysical price to pay for a solution to the problem of the past, namely the unreality of time. See Sprigge, especially for his critique of Whitehead's view of prehension and objectification (1983: 30–3, 225–32).

CHAPTER 7

1. In William James' chapter of *The Principles of Psychology* devoted to 'The stream of thought', he identifies five characters in thought, each of which agrees with Whitehead's generalisation of how actual occasions form a purely temporal nexus. Of these five, the last states that consciousness is 'interested in some parts of these objects to the exclusion of others, and welcomes or rejects – *chooses* from among them' (1891: 225, 284–90). This is the most important function of consciousness that is generalised in Whitehead's notion of positive and negative prehensions, and it is crucial for his explanation of how the actual occasion moulds the data provided by the past by selecting what is compatible with its subjective aim.

2. Stapp developed a quantum theory of consciousness that is based in part on the actual events ontology of Whitehead and Heisenberg and on the theory of consciousness advanced by James. See Stapp (1982, 2011).

3. In my paper on Whitehead's panpsychism (1995b), I argued that his view of the subjectivity of prehension committed him to a view that was roughly panpsychist. Much of the question of panpsychism, however, is largely semantic. Nevertheless, it should be clear that Whitehead certainly did not advance the view that all is mind (or consciousness) in the fashion of subjective or absolute idealists. For Whitehead, instead, it is experience that is basic. Some philosophers have identified this position as 'pan-experientialism' rather than panpsychism.

4. To be more precise, abiogenesis attempts to describe the process by which life arose from non-living matter at least 3.5 billion years ago, during the Eoarchean Era on planet earth. Biogenic graphite and microbial mat fossils provide the earliest evidence of the beginnings of life. As for the evolution of consciousness in general, there is nothing like a fossil record that could serve as a basis for estimating its appearance in the archaic sentient beings, but if we were to view the fourteen-billion-year evolution of the universe on the scale of one calendar year, human beings would appear to enter the picture on 31 December, very late in the evening. For theories of the origin of life, see John Casti's chapter 2, 'A warm little pond' (1990).

5. The view that relations are contained in experience was one of the main differences between Locke's and Hume's empiricism and James' radical empiricism. Whitehead's view of experience agrees with radical empiricism in that experience is understood to encompass more than the focus on the acquisition of clear and distinct sense data. It is rather phenomenologically dense and includes a sense of valuation and the passage of time. See James (1912: 39–122). Also see McHenry (1992: 78–87).

6. One exception to this is John Searle. Although he developed his argument along different lines than did Whitehead, Searle argues that when causality is viewed as already contained in the content of experience, we will not be looking for it as the object of perceptual experience and thereby will fail to find it. The experience itself is what does the causing (1983: 121–6).

7. Dorothy Emmet, however, notes a complication in the consistency of Whitehead's treatment of causality. In *Process and Reality*, Whitehead construed prehension in a backward-looking 'picking-up' view in contrast to a 'passing-on' view found in his earlier work, *Symbolism: Its Meaning and Effect*. I have focused more on the 'passing-on' view in the refutation of Hume, but recognise that the later 'picking-up' view largely results from Whitehead's conviction that we not only have perception *of* causation, but that perception *is* causation, since every causal event is an act of perception. See Emmet (1984: 170).

8. Also see Pierfrancesco Basile for an evaluation of Whitehead's critique of Hume on causation (2009: 63–80).

9. George Lucas, for example, thinks that in both Whitehead's and Parfit's view, 'personal identity is a function of sustained intentionality coordinated from moment to moment (and hence, both on-going task and achievement of individual agency) rather than a simple substantial reference to the location, properties, and causal activity of the physical body alone' (1989: 145–6). Also see Emmet (1992: 49–50).

10. For an overview of the bottom-up success of physical theory in a process of unification, see especially Stephen Hawking's 'Is the end in sight for theoretical physics?' (1993: 49–68). Steven Weinberg's distinction between 'hounds' and 'hawks', which is essentially a distinction between bottom-up and top-down approaches, has an affinity to Thomas Kuhn's distinction between 'normal science' and 'revolutionary science'. The former corresponds to the nuts-and-bolts working out of solutions to the puzzles within the established paradigm whereas the latter corresponds to the stratospheric, revolutionary theorising that creates the paradigm (1962).

References

Aristotle (1941), *Categories, Metaphysics* and *Physics. The Basic Works of Aristotle*, ed. Richard McKeon, New York: Random House.

Athearn, Daniel (1997), 'Whitehead as natural philosopher: anachronism or visionary?' *Process Studies* 26, 293–307.

Bardon, Adrian (2013), *A Brief History of the Philosophy of Time*, Oxford: Oxford University Press.

Barrett, Robert and Roger Gibson (eds) (1990), *Perspectives on Quine*, Oxford: Blackwell.

Barrow, John and Frank Tipler (1986), *The Anthropic Cosmological Principle*, Oxford: Clarendon Press.

Basile, Pierfrancesco (2009), *Leibniz, Whitehead and the Metaphysics of Causation*, Basingstoke: Palgrave Macmillan.

Bennett, Jonathan (1988), *Events and their Names*, Indianapolis: Hackett Publishing Company.

Berkeley, George ([1710] 1957), *A Treatise Concerning the Principles of Human Knowledge*, Indianapolis: Bobbs-Merrill.

Bohm, David (1980), *Wholeness and the Implicate Order*, London: ARK Paperbacks.

Bohm, David (1986), 'Time, the implicate order, and pre-space', *Physics and the Ultimate Significance of Time*, ed. David Ray Griffin, Albany: State University of New York Press, pp. 177–208.

Bohr, Niels (1934), *Atomic Physics and the Description of Nature*, Cambridge: Cambridge University Press.

Bohr, Niels (1958), *Atomic Physics and Human Knowledge*, New York: Wiley.

Broad, C. D. (1923), *Scientific Thought*, London: Routledge and Kegan Paul.

Broad, C. D. (1959), 'A reply to my critics', *The Philosophy of C. D. Broad*, ed. Paul Arthur Schilpp, New York: Tudor Publishing Company, pp. 709–830.

Bruno, Giordano (1584) *On the Infinite Universe and Worlds*, <www.positiveatheism.org/hist/brunoiuwo.htm> (last accessed 20 December 2014).

Burtt, E. A. (1953), 'Descriptive metaphysics', *Mind* LXXLL, 18–39.

Cahill, Reginald (2005), *Process Physics: From Information Theory to Quantum Space and Matter*, Hauppauge: Nova Science Publishers.

Campbell, Keith (1976), *Metaphysics*, Encino: Dickenson Publishing Company.

Čapek, Milič (1961), *The Philosophical Impact of Contemporary Physics*, New York: D. Van Nostrand Company.

Čapek, Milič (1991), *The New Aspects of Time: Its Continuity and Novelties*, Dordrecht: Kluwer Academic Publishers.

Carr, Bernard (ed.) (2007), *Universe or Multiverse?* Cambridge: Cambridge University Press.

Casti, John L. (1990), *Paradigms Lost*, New York: Avon Books.

Chew, Geoffrey F. (2004), 'A historical reality that includes big bang, free will, and elementary particles', *Physics and Whitehead: Quantum Process and Experience*, ed. Timothy Eastman and Hank Keeton, Albany: State University of New York Press, pp. 84–91.

Christian, William A. (1959), *An Interpretation of Whitehead's Metaphysics*, New Haven: Yale University Press.

Davidson, Donald (1980), *Essays on Actions and Events*, Oxford: Clarendon Press.

Davidson, Donald (1984), *Inquiries into Truth and Interpretation*, Oxford: Clarendon Press.

Davidson, Donald (1985), 'Reply to Quine on events', *Actions and Events: Perspectives on the Philosophy of Donald Davidson*, ed. Ernest LePore and Brian McLaughlin, Oxford: Blackwell, pp. 172–6.

Davies, Paul (2007), 'Universes galore: where will it all end?' *Universe or Multiverse?*, ed. Bernard Carr, Cambridge: Cambridge University Press, pp. 487–505.

Descartes, René ([1644] 1911), *Principles of Philosophy*, in *The Philosophical Works of Descartes*, trans. Elizabeth S. Haldane and G. R. T. Ross, Cambridge: Cambridge University Press.

Dummett, Michael (1994), *Origins of Analytical Philosophy*, Cambridge, MA: Harvard University Press.

Eastman, Timothy and Hank Keeton (eds) (2004), *Physics and Whitehead: Quantum Process and Experience*, Albany: State University of New York Press.

Edwards, Rem (2000), 'How process theology can affirm creation *ex nihilo*', *Process Studies* 29:1, 77–96.

Einstein, Albert (1923), 'On the electrodynamics of moving bodies', first pub. 1905, and 'Cosmological considerations on the general theory of relativity,' first pub. 1917, in *The Principle of Relativity*, H. A. Lorentz, A. Einstein, H. Weyl and H. Minkowski, trans. W. Perrett and G. B. Jeffrey, London: Methuen, pp. 35–65, 177–88.

Einstein, Albert (1948), 'Foreword', *The Universe and Dr. Einstein*, Lincoln Barnett, New York: Time Inc., pp. xviii–xix.

Einstein, Albert (1956), *Out of My Later Years*, New Jersey: The Citadel Press.

Einstein, Albert ([1920] 1962), *Relativity: The Special and General Theory*, London: Methuen.

Einstein, Albert (1982), 'Maxwell's influence on the development of the conception of physical reality', in Maxwell's *A Dynamical Theory of the*

Electromagnetic Field, ed. T. F. Torrance, Edinburgh: Scottish Academic Press, pp. 29–32.

Emmet, Dorothy (1984), 'Whitehead's view of causal efficacy', *Whitehead und der Prozessbegriff*, ed. Harald Holz and Ernest Wolf-Gazo, Freiburg and Munich: Verlag Karl Alber, pp. 161–78.

Emmet, Dorothy (1985), *The Effectiveness of Causes*, Albany: State University of New York Press.

Emmet, Dorothy (1992), *The Passage of Nature*, Philadelphia: Temple University Press.

Emmet, Dorothy (1996), 'Whitehead', *Philosophy* 71, 101–15.

Epperson, Michael (2004), *Quantum Mechanics and the Philosophy of Alfred North Whitehead*, New York: Fordham University Press.

Epperson, Michael and Elias Zafiris (2013), *Foundations of Relational Realism: A Topological Approach to Quantum Theory*, Lanham, MD: Lexington Books.

Fagg, Lawrence, W. (1997), 'Electromagnetism, time, and immanence in Whitehead's metaphysics', *Process Studies* 26, 308–17.

Feeney, Stephen, Matthew C. Johnson, Daniel J. Mortlock and Hiranya V. Peiris (2010), 'First observational tests of eternal inflation', Cosmology and Extragalactic Astrophysics (astro-ph.CO), <http://arxiv.org/abs/1012.1995> (last accessed 17 May 2014).

Feynman, Richard, Robert Leighton and Matthew Sands (1964), *The Feynman Lectures on Physics*, volumes I–III, Reading, MA: Addison-Wesley.

Folse, Henry J. (1974), 'The Copenhagen interpretation of quantum theory and Whitehead's philosophy of organism', *Tulane Studies in Philosophy* 23, 32–47.

Fritz, Charles A. (1952), *Bertrand Russell's Construction of the External World*, London: Routledge & Kegan Paul.

Gibson, Roger F., Jr (1982), *The Philosophy of W. V. Quine: An Expository Essay*, Tampa: University Presses of Florida.

Greene, Brian (2003), *The Elegant Universe*, New York: W. W. Norton & Company.

Greene, Brian (2004), *The Fabric of the Cosmos: Space, Time, and the Texture of Reality*, New York: Alfred A. Knopf.

Gribbin, John and Martin Rees (1989), *Cosmic Coincidences: Dark Matter, Mankind, and Anthropic Cosmology*, New York: Bantam Books.

Guth, Alan. H. (1981), 'Inflationary universe: a possible solution to the horizon and flatness problems', *Physical Review D* 23, 347–56.

Guth, Alan. H. (1997), *The Inflationary Universe: The Quest for a New Theory of Cosmic Origins*, Reading, MA: Helix Books.

Haack, Susan (1979), 'Descriptive and revisionary metaphysics', *Philosophical Studies* 35, 361–71.

Haack, Susan (1998), 'Between the Scylla of scientism and the Charybdis of apriorism', *The Philosophy of P. F. Strawson. The Library of Living*

Philosophers, volume XXVI, ed. Lewis Hahn, Chicago: Open Court, pp. 49–63.

Hacker, P. M. S. (1996), *Wittgenstein's Place in 20th-century Analytical Philosophy*, Oxford: Blackwell.

Hahn, Lewis (ed.) (1998), *The Philosophy of P. F. Strawson. The Library of Living Philosophers*, volume XXVI, Chicago: Open Court.

Hahn, Lewis and Paul Arthur Schilpp (eds) (1986), *The Philosophy of W. V. Quine. The Library of Living Philosophers*, volume XVIII, La Salle, IL: Open Court.

Harman, P. M. (2001), *The Natural Philosophy of James Clerk Maxwell*, Cambridge: Cambridge University Press.

Hättich, Frank (2004), *Quantum Processes: A Whiteheadian Interpretation of Quantum Field Theory*, Münster: Agenda Verlag.

Hawking, Stephen (1988), *A Brief History of Time: From the Big Bang to Black Holes*, New York: Bantam Books.

Hawking, Stephen (1993), *Black Holes and Baby Universes and Other Essays*, New York: Bantam Books.

Heil, John (2003), *From an Ontological Point of View*, Oxford: Clarendon Press.

Heil, John (2012), *The Universe as We Find It*, Oxford: Clarendon Press.

Heisenberg, Werner (1958), *Physics and Philosophy*, New York: Harper and Brothers.

Heitler, Walter (1945), *Elementary Wave Mechanics*, Oxford: Clarendon Press.

Holmgren, Christine and Leemon McHenry (2012), 'Quine and Whitehead on ontological reduction: properties reconsidered', *Process Studies* 41:2, 261–86.

Horgan, Terrence (1978), 'The case against events', *Philosophical Review* 87, 28–47.

Hubble, Edwin (1936), *The Realm of the Nebulae*, New Haven: Yale University Press.

Hume, David ([1748] 1927), *An Enquiry Concerning Human Understanding*, ed. L. A. Selby-Bigge, Oxford: Clarendon Press.

Hume, David ([1739] 1958), *A Treatise of Human Nature*, ed. L. A. Selby-Bigge, Oxford: Clarendon Press.

Hume, David ([1741, 1742] 1963), *Essays: Moral, Political and Literary*, Oxford: Oxford University Press.

Isham, C. J. (1993), 'Canonical quantum gravity and the problem of time', *Integrable Systems, Quantum Groups, and Quantum Field Theories*, ed. L. A. Ibort and M. A. Rodriguez, London: Kluwer Academic, pp. 157–288.

Isham, C. J. (1997), 'Structural issues in quantum gravity', *General Relativity and Gravitation: GR 14*, Singapore: World Scientific, pp. 167–209.

James, William (1891), *The Principles of Psychology*, volume I, London: Macmillan and Co.

James, William (1912), *Essays in Radical Empiricism*, New York: Longmans, Green and Co.

James, William ([1897] 1956), 'Is life worth living?' *The Will to Believe and Other Essays in Popular Philosophy*, New York: Dover Publications, pp. 32–62.

Johnson, Charles M. (1976), 'On prehending the past', *Process Studies* 6:4, 255–69.

Kneale, William (1949), *Probability and Induction*, Oxford: Clarendon Press.

Krips, Henry (1987), *The Metaphysics of Quantum Theory*, Oxford: Clarendon Press.

Kuhn, Thomas S. (1962), *The Structure of Scientific Revolutions*, Chicago: The University of Chicago Press.

Lango, John (2006), 'Whitehead's time through the prism of analytic concepts', *Les principes de la connaissance naturelle d'Alfred North Whitehead*, ed. Guillaume Durand and Michel Weber, Frankfurt: Ontos Verlag, pp. 137–56.

LePore, Ernest and Brian McLaughlin (eds) (1985), *Actions and Events: Perspectives on the Philosophy of Donald Davidson*, Oxford: Blackwell.

Lewis, C. I. (1923). 'A pragmatic conception of the a priori', *The Journal of Philosophy* 20, 169–77.

Locke, John ([1689] 1927), *An Essay Concerning Human Understanding*, Chicago: Open Court Publishing Company.

Lorentz, H. A., A. Einstein, H. Weyl, and H. Minkowski (1923), *The Principle of Relativity*, trans. W. Perrett and G. B. Jeffrey, London: Methuen.

Lowe, Victor (1966), *Understanding Whitehead*, Baltimore: Johns Hopkins University Press.

Lowe, Victor (1985) *Alfred North Whitehead: The Man and His Work, Volume I: 1861–1910*, Baltimore: The Johns Hopkins University Press.

Lowe, Victor (1990) *Alfred North Whitehead: The Man and His Work, Volume II: 1910–1947*, ed. J. B. Schneewind, Baltimore: The Johns Hopkins University Press.

Lucas, George R. (1989), *The Rehabilitation of Whitehead*, Albany: State University of New York Press.

McGinn, Colin (1999), *The Mysterious Flame: Conscious Minds in a Material World*, New York: Basic Books.

McHenry, Leemon (1990), 'Pan-physics: Whitehead's philosophy of natural science, 1918–1922', in Lowe, Victor, *Alfred North Whitehead: The Man and His Work, Volume II: 1910–1947*, ed. J. B. Schneewind, Baltimore: The Johns Hopkins University Press, pp. 108–30.

McHenry, Leemon (1992), *Whitehead and Bradley: A Comparative Analysis*, SUNY Series in Systematic Philosophy, Albany: State University of New York Press.

McHenry, Leemon (1995a), 'Quine's pragmatic ontology', *The Journal of Speculative Philosophy* 9, 147–58.

McHenry, Leemon (1995b), 'Whitehead's panpsychism as the subjectivity of prehension', *Process Studies* 24, 1–14.

McHenry, Leemon (1996), 'Descriptive and revisionary theories of events', *Process Studies* 25, 90–103.

McHenry, Leemon (1997), 'Quine and Whitehead: ontology and methodology', *Process Studies* 26, 2–12.

McHenry, Leemon (1998), 'Naturalized and pure metaphysics: a reply to Hutto', *Bradley Studies* 4, 97–101.

McHenry, Leemon (2000), 'The ontology of the past: Whitehead and Santayana', *The Journal of Speculative Philosophy* 14:3, 219–31.

McHenry, Leemon (2007) 'Maxwell's field and Whitehead's events: the adventure of a revolutionary idea', *Subjectivity, Process and Rationality*, ed. Michel Weber and Pierfrancesco Basile, Frankfurt: Ontos Verlag, pp. 177–189.

McHenry, Leemon (2008), 'Extension and the theory of the physical universe', *Handbook of Whiteheadian Process Thought*, volume I, ed. Michel Weber and William Desmond, Frankfurt: Ontos Verlag, pp. 291–302.

McHenry, Leemon (2011), 'The multiverse conjecture: Whitehead's cosmic epochs and contemporary cosmology', *Process Studies* 40:1, 5–25.

McTaggart, J. M. E. (1927), *The Nature of Existence*, volumes I and II, Cambridge: Cambridge University Press.

Malin, Shimon (2001), *Nature Loves to Hide: Quantum Physics and the Nature of Reality, A Western Perspective*, New York: Oxford University Press.

Maudlin, Tim (2010), 'Time, topology and physical geometry', *Proceedings of the Aristotelian Society Supplement*, LXXXIV, 63–78.

Maudlin, Tim (2012), *Philosophy of Physics: Space and Time*, Princeton: Princeton University Press.

Maxwell, James Clerk (1873), *A Treatise on Electricity and Magnetism*, Oxford: Clarendon Press.

Maxwell, James Clerk ([1865] 1982), *A Dynamical Theory of the Electromagnetic Field*, ed. T. F. Torrence, Edinburgh: Scottish Academic Press.

Maxwell, Nicholas (1985), 'Are probabilism and special relativity incompatible?' *Philosophy of Science* 52, 23–43.

Maxwell, Nicholas (1988), 'Quantum propensiton theory: a testable resolution of the wave/particle dilemma', *British Journal of Philosophy of Science* 39, 1–50.

Maxwell, Nicholas (1993), 'Induction and scientific realism: Einstein versus van Fraasen, part three: Einstein, aim-oriented empiricism and the discovery of special and general relativity', *British Journal of Philosophy of Science* 44, 275–305.

Maxwell, Nicholas (1994), 'Particle creation as the quantum condition for probabilistic events to occur', *Physics Letters A* 187, 351–5.

Maxwell, Nicholas (1998), *The Comprehensibility of the Universe*, Oxford: Clarendon Press.

Maxwell, Nicholas (2006), 'Special relativity, time, probabilism, and ultimate reality', *The Ontology of Space-Time*, ed. D. Dieks, Amsterdam: Elsevier, pp. 229–45.

Mei, Tsu-Lin (1961), 'Subject and predicate, a grammatical preliminary', *The Philosophical Review* VXX, 153–75.

Minkowski, Herman (1923), 'Space and time', *The Principle of Relativity*, H. A. Lorentz, A. Einstein, H. Weyl and H. Minkowski, trans. W. Perrett and G. B. Jeffrey, London: Methuen, pp. 75–91.

Mulvaney, Robert J. and Philip M. Zeltner (eds) (1981), *Pragmatism: Its Sources and Prospects*, Columbia, SC: University of South Carolina Press.

Nagel, Ernest (1946), 'Russell's philosophy of science', *The Philosophy of Bertrand Russell. The Library of Living Philosophers*, volume V, ed. Paul Arthur Schilpp, Menasha: George Banta Publishing Company, pp. 319–49.

Newton, Isaac (1962), '*De gravitatio et aequipondio fluidorum*', *Unpublished Scientific Papers of Isaac Newton*, ed. A. R. Hall and M. Boas Hall, Cambridge: Cambridge University Press, pp. 121–56.

Newton, Isaac ([1687] 1995), *The Principia*, trans. A. Motte and F. Cajori (1729), New York: Prometheus Books.

Nobo, Jorge (1986), *Whitehead's Metaphysics of Extension and Solidarity*, Albany: State University of New York Press.

Norton, John D. (2014), 'The hole argument', *The Stanford Encyclopedia of Philosophy*, ed. Edward N. Zalta, <http: //plato.stanford.edu/archives/spr2014/entries/spacetime- holearg/> (last accessed 30 July 2014).

Papatheodorou, C. and Basil Hiley (1997), 'Process, temporality and space-time', *Process Studies* 26, 247–78.

Peebles, P. J. E. (1971), *Physical Cosmology*, Princeton: Princeton University Press.

Plato (1953), *Theaetetus* and *Timaeus*, trans. Benjamin Jowett, volume III, Oxford: Clarendon Press.

Popper, Karl (1959), *The Logic of Scientific Discovery*, New York: Basic Books.

Popper, Karl (1992), *The Postscript to The Logic of Scientific Discovery*, volume 3, *Quantum Theory and the Schism in Physics*, London: Routledge.

Quine, W. V. (1934), *A System of Logistic*, Cambridge, MA: Harvard University Press.

Quine, W. V. (1941), 'Whitehead and the rise of modern logic', *The Philosophy of Alfred North Whitehead, The Library of Living Philosophers*, volume III, ed. Paul Arthur Schilpp, New York: Tudor Publishing Company, pp. 125–63.

Quine, W. V. (1953), *From a Logical Point of View*, Cambridge, MA: Harvard University Press.

Quine, W. V. (1960), *Word and Object*, Cambridge, MA: The MIT Press.

Quine, W. V. (1969), *Ontological Relativity and Other Essays*, New York: Columbia University Press.

Quine, W. V. (1970), *Philosophy of Logic*, Cambridge, MA: Harvard University Press.

Quine, W. V. (1976), 'Whither physical objects?' *Essays in Memory of Imre*

Lakatos, *Boston Studies in the Philosophy of Science*, ed. R. S. Cohen et al., Dordrecht: D. Reidel, pp. 497–504.

Quine, W. V. (1981a), 'The pragmatists' place in empiricism', *Pragmatism: Its Sources and Prospects*, ed. Robert J. Mulvaney and Philip M. Zeltner, Columbia, SC: University of South Carolina Press.

Quine, W. V. (1981b), *Theories and Things*, Cambridge, MA: Harvard University Press.

Quine, W. V. (1985a), 'Events and reification', *Actions and Events: Perspectives on the Philosophy of Donald Davidson*, ed. Ernest LePore and Brian McLaughlin, Oxford: Blackwell, pp. 162–71.

Quine, W. V. (1985b), *The Time of My Life: An Autobiography*, Cambridge, MA: The MIT Press.

Quine, W. V. (1986), 'Autobiography of W. V. Quine', *The Philosophy of W. V. Quine. The Library of Living Philosophers*, volume XVIII, ed. Lewis Hahn and Paul Arthur Schilpp, La Salle, IL: Open Court.

Quine, W. V. (1997), 'Response to Leemon McHenry', *Process Studies* 26, 13–14.

Quine, W. V. (2008a), *Quine in Dialogue*, ed. Dagfinn Føllesdal and Douglas Quine, Cambridge, MA: Harvard University Press.

Quine, W. V. (2008b), *Confessions of a Confirmed Extensionalist and Other Essays*, ed. Dagfinn Føllesdal and Douglas Quine, Cambridge, MA: Harvard University Press.

Quinton, Anthony (1985), 'The right stuff', *The New York Review of Books* 32:19, 52.

Redhead, Michael (1995), *From Physics to Metaphysics*, Cambridge: Cambridge University Press.

Rees, Martin (1997), *Before the Beginning: Our Universe and Others*, Reading, MA: Helix Books.

Rees, Martin (2000), *Just Six Numbers: The Deep Forces that Shape the Universe*, New York: Basic Books.

Rees, Martin (2001), *Our Cosmic Habitat*, Princeton: Princeton University Press.

Rees, Martin (2007), 'Cosmology and the multiverse', *Universe or Multiverse?* ed. Bernard Carr, Cambridge: Cambridge University Press, pp. 57–75.

Rorty, Richard (1979), *Philosophy and the Mirror of Nature*, Princeton: Princeton University Press.

Russell, Bertrand (1921), *The Analysis of Mind*, London: George, Allen & Unwin.

Russell, Bertrand ([1927] 1934), *The Analysis of Matter*, London: Kegan Paul.

Russell, Bertrand (1945), *A History of Western Philosophy*, New York: Simon and Schuster.

Russell, Bertrand (1946), 'Replies to criticisms', *The Philosophy of Bertrand Russell. The Library of Living Philosophers*, volume V, ed. Paul Arthur Schilpp, Menasha: George Banta Publishing Company, pp. 681–741.

Russell, Bertrand ([1914] 1956), *Our Knowledge of the External World*, New York: Mentor Books.

Russell, Bertrand (1963), *Portraits from Memory and Other Essays*, New York: Simon and Schuster.

Russell, Bertrand ([1959] 1995), *My Philosophical Development*, London: Routledge.

Santayana, George (1955), *Scepticism and Animal Faith*, New York: Dover.

Schilpp, Paul Arthur (ed.) (1941), *The Philosophy of Alfred North Whitehead. The Library of Living Philosophers*, volume III, New York: Tudor Publishing Company.

Schilpp, Paul Arthur (ed.) (1946), *The Philosophy of Bertrand Russell. The Library of Living Philosophers*, volume V, Menasha: George Banta Publishing Company.

Schilpp, Paul Arthur (ed.) (1959), *The Philosophy of C. D. Broad. The Library of Living Philosophers*, volume X, New York: Tudor Publishing Company.

Searle, John (1983), *Intentionality*, Cambridge: Cambridge University Press.

Sherburne, Donald ([1966] 1981), *A Key to Whitehead's Process and Reality*, Chicago: The University of Chicago Press.

Shields, George W. (ed.) (2003), *Process and Analysis: Whitehead, Hartshorne, and the Analytic Tradition*, Albany: State University of New York Press.

Shimony, Abner (1965), 'Quantum physics and the philosophy of Whitehead', *Boston Studies In the Philosophy of Science*, volume II, ed. Robert Cohen and Max W. Wartofsky, New York: Humanities Press, pp. 307–42.

Sider, Theodore (2001), *Four-Dimensionalism: An Ontology of Persistence and Time*, Oxford: Oxford University Press.

Simons, Peter (2003), 'Review of Shields', *Transactions of the Charles Peirce Society* 39:4, 663–6.

Smolin, Lee (1997), *The Life of the Cosmos*, New York: Oxford University Press.

Sprigge, T. L. S. (1983), *The Vindication of Absolute Idealism*, Edinburgh: Edinburgh University Press.

Sprigge, T. L. S. (1994), 'Consciousness', *Synthese* 98, 73–93.

Stapp, Henry (1972), 'The Copenhagen interpretation', *American Journal of Physics* 40, 1098–116.

Stapp, Henry (1979), 'Whiteheadian approach to quantum theory and the generalized Bell's Theorem', *Foundations of Physics* 9, 1–25.

Stapp, Henry (1982), 'Mind, matter and quantum mechanics', *Foundations of Physics* 12, 363–99.

Stapp, Henry (2009), 'Quantum collapse and the emergence of actuality from potentiality', *Process Studies* 38, 319–39.

Stapp, Henry (2011), *Mindful Universe: Quantum Mechanics and the Participating Observer*, Berlin: Springer.

Strawson, P. F. (1959), *Individuals: An Essay in Descriptive Metaphysics*, London: Methuen.

Strawson, P. F. (1990), 'Two conceptions of philosophy', *Perspectives on*

Quine, ed. Robert Barrett and Roger Gibson, Oxford: Blackwell, pp. 310–18.

Strawson, P. F. (1992), *Analysis and Metaphysics: An Introduction to Philosophy*, Oxford: Oxford University Press.

Strawson, P. F. (1998), 'Reply to Susan Haack', *The Philosophy of P. F. Strawson. The Library of Living Philosophers*, volume XXVI, ed. Lewis Hahn, Chicago: Open Court, pp. 64–7.

Tegmark, Max (2007), 'The multiverse hierarchy', *Universe or Multiverse?*, ed. Bernard Carr, Cambridge: Cambridge University Press, pp. 99–125.

Thalberg, Irving (1985), 'A world without events?' *Essays on Davidson: Actions and Events*, ed. Bruce Vermazen and Merrill Hintikka, Oxford: Clarendon Press, pp. 137–55.

Tolman, R. C. (1934), *Relativity, Thermodynamics and Cosmology*, Oxford: Clarendon Press.

Unwin, Nicholas (1996), 'The individuation of events', *Mind* 105, 315–30.

Vermazen, Bruce and Merrill Hintikka (eds) (1985), *Essays on Davidson: Actions and Events*, Oxford: Clarendon Press.

Weinberg, Steven (1988), *The First Three Minutes*, New York: Basic Books.

Weinberg, Steven (1994), *Dreams of a Final Theory*, New York: Vintage Books.

Whitehead, Alfred North (1919), *An Enquiry Concerning the Principles of Natural Knowledge*, Cambridge: Cambridge University Press.

Whitehead, Alfred North (1920), *The Concept of Nature*, Cambridge: Cambridge University Press.

Whitehead, Alfred North (1922), *The Principle of Relativity with Applications to Physical Science*, Cambridge: Cambridge University Press.

Whitehead, Alfred North (1925), *Science and the Modern World*, Cambridge: Cambridge University Press.

Whitehead, Alfred North (1927), *Symbolism: Its Meaning and Effect*, New York: Macmillan Company.

Whitehead, Alfred North (1933), *Adventures of Ideas*, Cambridge: Cambridge University Press.

Whitehead, Alfred North ([1929] 1958), *The Function of Reason*, Boston, MA: Beacon Press.

Whitehead, Alfred North ([1938] 1968), *Modes of Thought*, New York: The Free Press.

Whitehead, Alfred North ([1947] 1974), *Science and Philosophy*, New York: Philosophical Library.

Whitehead, Alfred North ([1929] 1978), *Process and Reality: An Essay in Cosmology*, corrected edition, ed. David Ray Griffin and Donald W. Sherburne, New York: The Free Press.

Whitehead, Alfred North ([1926] 1996), *Religion in the Making*, New York: Fordham University Press.

Whitehead, Alfred North and Bertrand Russell (1910–1913), *Principia Mathematica*, volumes I–III, Cambridge: Cambridge University Press.

Whorf, Benjamin L. ([1956] 1970), *Language, Thought and Reality*, ed. John B. Carroll, Cambridge, MA: The MIT Press.

Wolszczan, Aleksander and Dale Frail (1992), 'A planetary system around the millisecond pulsar PSR 1257+12', *Nature* 355:6356, 145–7.

Yourgrau, Palle (1991), *The Disappearance of Time: Kurt Gödel and the Idealistic Tradition in Philosophy*, Cambridge: Cambridge University Press.

Index